Applying Innovation

Applying Innovation

David O'Sullivan
National University of Ireland, Galway

Lawrence Dooley
University College Cork, Ireland

Los Angeles • London • New Delhi • Singapore

For information:

SAGE Publications, Inc.
2455 Teller Road
Thousand Oaks, California 91320
E-mail: order@sagepub.com

SAGE Publications India Pvt. Ltd.
B 1/I 1 Mohan Cooperative
   Industrial Area
Mathura Road, New Delhi 110 044
India

SAGE Publications Ltd.
1 Oliver's Yard
55 City Road
London EC1Y 1SP
United Kingdom

SAGE Publications Asia-Pacific Pte. Ltd.
33 Pekin Street #02-01
Far East Square
Singapore 048763

Printed in the United States of America

Library of Congress Cataloging-in-Publication Data

O'Sullivan, David.
Applying innovation/David O'Sullivan, Lawrence Dooley.
     p. cm.
Includes bibliographical references and index.
ISBN 978-1-4129-5454-9 (cloth)
ISBN 978-1-4129-5455-6 (pbk.)
    1. Technological innovations—Management. 2. Organizational change.
I. Dooley, Lawrence. II. Title.

HD45.O88 2009
658.4′063—dc22   2008006237

This book is printed on acid-free paper.

08   09   10   11   12   10   9   8   7   6   5   4   3   2   1

Acquisitions Editor:	Al Bruckner
Editorial Assistant:	MaryAnn Vail
Production Editor:	Diane S. Foster
Copy Editor:	Carol Anne Peschke
Typesetter:	C&M Digitals (P) Ltd.
Proofreader:	Scott Oney
Indexer:	Diggs Publication Services
Cover Designer:	Gail Buschman
Marketing Manager:	Nichole M. Angress

Brief Contents

Detailed Contents

Introduction

Innovation is an important force in creating and sustaining organizational growth. Effective innovation can mean the difference between leading with a particular product, process, or service and simply following the pack, with the resulting risk of stagnation and decline. Innovation transforms mediocre companies into world leaders and ordinary organizations into stimulating environments for employees. Innovation is the process of making changes to something established by introducing something new; these changes can be either radical or incremental. All organizations need to innovate, whether they are profit or nonprofit. Innovation is as relevant to services in a public hospital as it is for products and processes in a manufacturing company. Innovation takes place throughout an organization, from management boards and individual departments to project teams and individuals. In today's global economy innovation is often a collaborative activity that takes place across extended organizations and includes suppliers, distributors, and other strategic alliances. Despite its importance, many organizations fail to recognize the need for innovation and to develop skills to innovate on a continuous basis. If an organization is to be sustainable, it must develop its capability to manage its innovation process.

The term *innovation* is ambiguous to some and often associated with visions of organizations that can create world-beating products that grow to dominate entire sectors of industry. This view is informing and often entertaining, but it also allows many practitioners to shy away from engaging in the concept of innovation, in the knowledge that such visions are rarely realized and often depend on factors such as previous market dominance and chance. This book is about looking into the practical techniques, large and small, practiced every day in leading organizations, that are used to manage innovation. Whereas innovation theory can inform the decisions that must be made by organizations, this book is primarily about the tools and techniques that put structure on the decisions organizations must make for themselves. Although the decisions vary significantly between organizations, the structures around the management of innovation are essentially the same.

Applying Innovation is about describing how to systematically deliver innovations that add value to customers. The approach adopted is a symbiosis of management techniques that include innovation management, strategic planning, performance measurement, creativity, project portfolio management, performance appraisal, and knowledge management. Whereas other books offer in-depth insights into one or more of these areas, this book provides ingredients from each that combine into an easy-to-understand framework for applying innovation in any organization. This book contains a systematic approach to creating structure for the application of innovation in any organization. Applying innovation requires close attention to five key types of knowledge—goals, actions, teams, results, and communities—and, perhaps of equal importance, the relationships among all five. This book will show you how to develop a simple knowledge management process by which people anywhere in an organization can share innovation-related information. Systematic structuring and sharing of information can dramatically improve an organization's quest to add more value to customers, and that in turn helps organizations grow.

Book Aims

One of the main aims of this book is to map out the main concepts of the innovation process into an easy-to-understand and easy-to-apply framework: the innovation funnel. This book will help you understand key concepts in innovation management and particularly how to apply innovation to any organization. Innovation is defined as "making changes, large and small, to products, processes, and services in any organization, profit or nonprofit, that add value to customers and continuing to learn from that process so that it can be repeated continuously." The aims of this book stem from this definition: to facilitate the learning and practicing of techniques, tools, and methods for applying innovation management. For this reason the book does not focus heavily on innovation theory but rather focuses on tools and techniques that have gained widespread acceptance in leading innovative organizations. This book is developed around a number of specific learning outcomes:

- Understanding key concepts in the theory and process of innovation
- Understanding how to manage and apply innovation
- Using explicit skills for defining goals, generating ideas, empowering teams, and monitoring the results of innovation
- Developing a simple knowledge management system for managing innovation
- Working effectively as an individual and as a member of a team

- Presenting, communicating, and promoting innovation plans
- Applying what you have learned to managing innovation in any organization

Book Structure

The book is structured around the key concepts of applying innovation that are presented as elements of an easy-to-understand framework: the innovation funnel. For this reason this book is divided into five parts. The first part introduces innovation and, in particular, innovation management. The next four parts cover the four interrelated areas of applying innovation: goals, actions, teams, and results.

The five parts are as follows:

Part I: Understanding Innovation

Part II: Defining Innovation Goals

Part III: Managing Innovation Actions

Part IV: Empowering Innovation Teams

Part V: Sharing Innovation Results

Part I, "Understanding Innovation," describes the main concepts behind the innovation process. Innovation is classified according to its impact on products, processes, and services. The difference between radical and incremental innovation is discussed. The special relationship between product innovation and process innovation is introduced. The process of managing innovation is then described. This part concludes with a chapter describing how the innovation management process can be structured and put to work toward achieving the objectives of the organization. The concept of the innovation funnel is presented as a metaphor for understanding innovation management in any organization.

Part II, "Defining Innovation Goals," describes the process of setting goals for innovation. Various goal-setting activities are described. Defining statements such as mission and vision statements are outlined. Identifying the key stakeholders for an organization is discussed, particularly in terms of what they require from the organization. Creating a set of strategic objectives for innovation is then described. This part concludes with a chapter that outlines how to create performance indicators for measuring the impact of the innovation process.

Part III, "Managing Innovation Actions," describes the various ways in which innovation goals can be achieved. Problem solving is often regarded as the first step in the innovation process, where problems are identified and solved systematically. Generating ideas that solve problems or meet particular goals is an important part of the innovation process. Various techniques for idea generation are described. Many ideas grow to become large-scale initiatives or projects that must be managed. A special kind of project is one that involves radical new product development. The final chapter in Part III looks at how to manage and balance a portfolio of projects.

Part IV, "Empowering Innovation Teams," describes a number of ways in which people can work together more effectively to achieve innovation goals. The part begins with a chapter that looks at leadership and individual responsibility. The concept of teams is then outlined, as is the importance of team participation in achieving goals. Motivation and reward are an important part of the innovation and creativity process. A simple but effective performance appraisal system is outlined by which people can be aligned with the innovation goals of the organization.

Part V, "Sharing Innovation Results," describes how to monitor and share information related to the goals, actions, and teams in the innovation process. The part begins with a chapter on knowledge management and how knowledge can be codified and shared in a collaborative environment. The next chapter looks at developing a simple knowledge management system that can later be expanded. A critical part of monitoring the results of innovation and making decisions is mapping the relationships between goals and actions, between teams and goals, and between actions and teams. These relationships can be described using a simple mapping technique. The part concludes with a chapter that describes how to extend innovation across an organization that includes management boards, departments, project teams, suppliers, distributors, and other strategic partners. Each of these organizations shares a common goal of providing value to customers, both internal and external.

The relationships between the key knowledge areas discussed in these parts are illustrated in Figure 0.1. This figure is called the innovation funnel, and it forms the framework for understanding the knowledge associated with the application of innovation in any organization.

The innovation funnel is a metaphor for understanding how to apply innovation and a structured way of defining the information requirements for managing and sharing innovation-related knowledge. The funnel illustrates how goals, actions, teams, and results interact with each other to manage innovation. Actions enter the mouth of the funnel and represent, among other things, alternative ideas for innovation. These ideas can come from many sources, including lead users and employees. These actions flow to the neck of the funnel, where they are evaluated and filtered. In the filtering process some will be eliminated, merged, or delayed and others processed into projects or initiatives that make change happen. This filtering

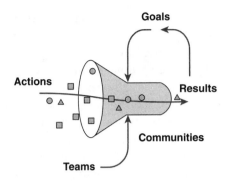

Figure 0.1   Innovation Funnel

process is controlled and constrained mainly by innovation goals and innovation teams. These controls loosen or tighten the neck of the funnel. Tightly defined goals can be visualized as closing the neck of the funnel, allowing fewer ideas through. This can simultaneously encourage the creation or identification of new ideas that have a stronger relationship with goals. On the other hand, better engagement of individuals and teams can be visualized as opening the neck of the funnel, allowing more ideas to be resourced and developed. The final arrow shown in the funnel is results. Results flow from the narrow end of the funnel and represent information about the progress of goals, actions, and teams. This arrow flows back toward goals, representing the impact of results on the process of defining and redefining innovation goals that continuously guide the innovation process.

The management and execution of goals, actions, teams, and results takes place in various communities within an organization where information can be easily communicated and shared and where knowledge can be managed effectively. These communities often take the form of management groups, departments, project teams, and even individuals but can also extend beyond traditional boundaries to include suppliers, distributors, and even customers. Every organization can have many such communities and consequently many innovation funnels. Applying innovation is concerned not only with how information is structured and shared within each funnel but also with how innovation information is shared between funnels across the extended organization.

Learning Activities

This book is designed to help you learn the fundamentals of applying innovation. In addition to the information given, there are a number of activities in each chapter to help you apply some of the ideas presented. These activities enable you to relate the information covered in the different chapters to your own organization. Your organization can be based on a real organization, or it can be created from your imagination and experience. You are encouraged to create an innovation plan for this organization— your own detailed case study. The activities will combine into one dynamic innovation plan for this organization. There is a sample innovation plan in the Appendix to help you in the various decisions you will have to make. Your innovation plan can be created using some simple spreadsheet templates used in the book, or you may decide to build your own online knowledge management system from one of the many organizations that provide this service. Each chapter also contains questions that are designed to help you reflect on some of the main points raised in the chapter. You are encouraged to reflect on the answers to these questions as part of your learning process. Together, the activities and reflection questions provide a way for you to embed core knowledge in your memory and apply it to building and operating an innovation process.

Intended Audience

This book is designed for those pursuing upper-level undergraduate degrees and master's degrees in business, science, and engineering. The book is also a practical accompaniment to courses that use books dealing mainly with the theory of change and theory of innovation management. This book is also suitable for managers, team leaders, and individuals in for-profit and nonprofit organizations. The information and approach adopted by this book are relevant and applicable to all types of organizations: manufacturing, service, healthcare, public service, nongovernment, and so on. Instructors will find the book an important resource for teaching innovation-related topics such as strategic planning, performance measurement, teams and leadership, knowledge management, and project management.

Key Terms

Throughout this book, a number of key terms are used repeatedly. Understanding and applying a specific term can be the key to unlocking meaning and relationships between concepts. Many terms have synonyms that may be preferred in other texts. For example, the terms *project* and *indicator* are used in this book, whereas other texts may use the terms *initiative* and *measure*, respectively. Practice your understanding of the specific terms presented in Table 0.1.

Table 0.1   Common Terms Used in This Book

Term Used in Book	Brief Explanation
Action	Expenditure of effort
Change	Process through which something becomes different
Communities	Groups of individuals with a common purpose
Customer	Individual or organization that requires product, process, or service
Deliverable	Unit of output of an effort
Effort	Vigorous or determined attempt to do something
Goal	Objective of an effort
Idea	Effort in generating a new concept
Indicators	Measurements or metrics of specific goals for an effort
Individuals	Human resources for an effort
Innovate	To make changes to something established
Innovation	Process of making changes through effort
Knowledge	Fact, information, or skills acquired through learning or experience
Learning	Gaining or acquiring knowledge
Milestone	Point or stage on a timeline where effort can be reviewed
Mission	Important assignment
Objective	Statement of desired end point for an effort
Organization	Systematic arrangement of individuals with a particular mission
Portfolio	Group of projects
Problem	Idea for existing or future failures
Process	Series of actions to achieve an end
Product	Article or substance processed for sale
Project	New initiative
Requirements	Statements of desired functionality for an effort
Resources	Individuals, teams, equipment, or money needed for an effort
Responsible	Assigned to lead or account for an effort
Review	Performance appraisal of an individual for an effort
Service	Action of doing work for an individual or organization
Skill	Ability to perform an effort well
Statement	Declared message of an objective or position in an effort
Task	A subset of a workpackage or project
Team	Group of individuals or resources needed for an effort
Workpackage	A group of tasks in one project

Part I

Understanding Innovation

All organizations need to change in order to sustain current activities and develop growth. Some changes are positive and lead to increased growth in terms of efficiency, quality, or revenue; other changes are negative and risk stagnation and decline for the organization. The principal mechanism for change in any organization is innovation. Innovation is the process of making changes to something established by introducing something new that adds value to customers. This part describes the main concepts behind innovation and its impact on products, processes, and services. The difference between radical and incremental innovation is discussed, including the special case of disruptive innovation. The relationship between innovation and change is explored. We then look at investment that organizations make in innovation and the goals they endeavor to achieve and also the reasons why many innovations do not achieve these goals. We also examine the organizational cultural factors that can nurture and support the innovation effort. This part concludes by looking at an effective process of applying innovation and describes how the innovation management process can be structured and put to work to help organizations to grow continuously. The concept of the innovation funnel is presented as a powerful metaphor for understanding the application of innovation in any organization.

LEARNING TARGETS

When you have completed this part you will be able to

- Develop an understanding for organizational innovation
- Explain the difference between product, process, and service innovation
- Understand radical and incremental innovation
- Understand the reasons why organizations invest in innovation
- Explain the main causes of failure in innovation
- Understand a process for managing innovation
- Describe the importance of the innovation funnel metaphor

Defining Innovation 1

Innovation is about helping organizations grow. Growth is often measured in terms of turnover and profit, but can also occur in knowledge, in human experience, and in efficiency and quality. Innovation is the process of making changes to something established by introducing something new. As such, it can be radical or incremental, and it can be applied to products, processes, or services and in any organization. It can happen at all levels in an organization, from management teams to departments and even to the level of the individual.

This chapter describes the main concepts behind innovation. We explore the different types of innovation that affect the growth of an organization. The difference between radical and incremental innovation is discussed, as is the special relationship between product and process innovation.

LEARNING TARGETS

When you have completed this chapter you will be able to

- Define innovation and explain the difference with related terms
- Understand the drivers of the need for innovation and change
- Explain product, process, and service innovation
- Describe the difference between radical and incremental innovation
- Define disruptive technology
- Show how product and process innovations are related
- Explain the relationship between innovation and operations

Definition of Innovation

Innovation has been and continues to be an important topic of study for a number of different disciplines, including economics, business, engineering, science, and sociology. Despite the fact that innovation has been studied in a variety of disciplines, the term is often poorly understood and can be sometimes confused with related terms such as *change, invention, design,* and *creativity.* Most people can provide examples of innovative products such as the iPod or the PC, but few can clearly define the innovative aspects of these products. Among academics there is a difference of opinion about what the term *innovation* really means. One definition of *innovation* taken from the dictionary that fits the ideas and concepts used in this book is the following (*The New Oxford Dictionary of English,* 1998, p. 942):

> *Making changes to something established by introducing something new.*

This definition does not suggest that innovation must be radical or that it occurs exclusively to products. Nor does it suggest that innovation is exclusively for large organizations or single entrepreneurs. Nor does it suggest that it is exclusively for profit-making businesses; innovation is as relevant for a hospital or local government as it is for a business. In the organizational context innovation can occur to products, processes, or services. It can be incremental or radical, and it can occur at various levels in an organization, from management groups and departments to project teams and even individuals.

This is the general concept of innovation as discussed in this book. We will see later that the fundamental concepts of innovation as they are derived from this definition are universally relevant for all organizations, from private companies such as Nokia down to public organizations such as hospitals. Innovation is a process that transforms ideas into outputs, which increase customer value. The process can be fed by both good and bad ideas. In management of the innovation process, destroying poor ideas often is as important as nurturing good ones; in this way, scarce resources can be released and good ideas spotlighted. Every good idea usually replaces an older established one. The goal of every organization is the successful development of good ideas. To express this development of good ideas in innovation, we need to add an addendum to our definition:

> *Innovation is the process of making changes to something established by introducing something new that adds value to customers.*

This addendum is important. By describing an innovation as adding value to customers, we assume naturally that customers who experience the added value will continue to use the product, process, or service or at least have an improved experience. This in turn will lead to growth for the organization. Innovation management is the process of managing innovation within an organization. This includes activities such as managing ideas, defining goals, prioritizing projects, improving communications, and motivating teams. As we will see later, innovations have particular life cycles; today's innovation will become obsolete in the future. For organizations to sustain their mission, they must continuously innovate and replace existing products, processes, and services with more effective ones. Focusing on innovation as a continuous process acknowledges the effect that learning has on knowledge creation within the organization. Learning how to innovate effectively entails managing knowledge within the organization and offers the potential to enhance the way the organization innovates. This element adds a further extension to our definition:

> *Innovation is the process of making changes to something established by introducing something new that adds value to customers and contributes to the knowledge store of the organization.*

The concept of an organization's knowledge store is partially synonymous with the concept of organizational learning. An organization that can continuously learn and adapt its behavior to external stimuli does so by continuously adding to its collective knowledge store. The emerging perspective by specialists in the field of innovation is to define innovation in the broadest context possible. One reason for this is that if it is defined too narrowly, it may limit creativity by excluding certain avenues of investigation. Innovation is linked to the concepts of novelty and originality. However, novelty is highly subjective. What may be a trivial change for one organization may be a significant innovation for another. Based on this perspective, we can further extend the definition of innovation as follows:

> *Innovation is the process of making changes, large and small, radical and incremental, to products, processes, and services that results in the introduction of something new for the organization that adds value to customers and contributes to the knowledge store of the organization.*

This latter definition, although general, is specific enough to illustrate a number of core concepts of innovation as applied in any organization. Applying innovation, which is the main focus of this book, can be defined by adding a number of key words to the preceding definition.

> *Applying innovation is the application of practical tools and techniques that make changes, large and small, to products, processes, and services that results in the introduction of something new for the organization that adds value to customers and contributes to the knowledge store of the organization.*

Related Concepts

Innovation is often used in conjunction with terms such as *creativity, design, invention,* and *exploitation.* It is also closely associated with terms such as *growth* and *change.* Let's explore these relationships in more detail in order to get a deeper understanding of what we mean by *innovation.* Related concepts include invention, growth, creativity, design, exploitation, change, failure, entrepreneurship, customers, knowledge, and society.

INNOVATION AND INVENTION

Invention is a term often used in the context of innovation. *Invention* has its own separate entry in the dictionary and is defined as follows (*The New Oxford Dictionary of English,* 1998, p. 960):

> *Creating something new that has never existed before.*

Invention need not fulfill any useful customer need and need not include the exploitation of the concept in the marketplace. Innovation differs from invention in that it is more than the creation of something novel; it also includes the exploitation for benefit by adding value to customers. Invention is often measured as the ability to patent an idea. If this can be achieved, then it is an invention. The success or failure of an invention depends not only on the ideas chosen by the organization but also on how well their implementation is managed. Invention is often about creating something that has yet to be desired by a customer. Numerous inventions never lead to innovation because they are never brought to the marketplace. If an invention can be exploited and transformed into change that adds value to a customer, then it becomes an innovation. On the other hand, there are many innovations that do not require invention in terms of originality. Process and service innovations often involve applying well-established techniques and technology. Although it can be argued that this does not encompass invention because it already exists, it is still a legitimate form of innovation because it is novel to the organization applying it.

INNOVATION AND GROWTH

Innovation is about developing growth. According to Drucker (1988), innovation can be viewed as a purposeful and focused effort to achieve change in (an organization's) economic or social potential. Bottom-line growth can occur in a number of ways, such as better service quality and shorter lead times in nonprofit organizations and cost reduction, cost avoidance, and increased turnover in profit-focused organizations.

INNOVATION AND CREATIVITY

Creativity is regarded as a key building block for innovation (Rosenfeld & Servo, 1991) and is an inherent capability in all human beings. Creativity is a mental process that results in the production of novel ideas and concepts that are appropriate, useful, and actionable. The creative process can be said to consist of four distinct phases: preparation, incubation, illumination, and verification (Wallas, 1926). Later revisions of this process have added a final phase, elaboration (Kao, 1989), in which the idea is structured and finalized in a form that can be readily communicated to others. Creativity entails a level of originality and novelty that is essential for innovation. Although creativity is a fundamental part of innovation, it is wrong to interchange the terms. Innovation encourages the further processing of the output of the creative process (the idea) so as to allow the exploitation of its potential value through development.

INNOVATION AND DESIGN

The term *design* in the context of innovation is defined as "the conscious decision-making process by which information (an idea) is transformed into an outcome be it tangible (product) or intangible (service)" (von Stamm, 2003, p. 11). The design activity draws heavily on creativity to resolve issues such as the aesthetics, form, and functionality of the eventual outcome. In this way, during the exploitation phase of the innovation process, organizations engage in design activities that will produce an output that provides the optimum fit with market requirements. Although design is an integral part of the exploitation phase of an innovation, it is only one aspect. Exploitation can include other elements, such as process development and market preparation.

INNOVATION AND EXPLOITATION

There are numerous alternative definitions of *innovation*. One popular alternative is to present innovation as an invention that has been exploited commercially (Martin, 1994). In this alternative definition, the term *invention* has the same meaning as mentioned earlier, that is, something new that has never existed before. This creation of something new derives from the creative capability of the organization and provides opportunities to be exploited. This alternative definition of innovation has been expressed as follows (Roberts, 1988):

$$Innovation = Invention + Exploitation$$

Therefore, innovation can be viewed as the systematic approach to creating an environment based on creative discovery, invention, and commercial exploitation of ideas that meet unmet needs (Bacon & Butler, 1998). This definition fits in very well with many high-profile examples of innovation, such as the invention of the transistor used in computers or radio-frequency identity (RFID) tags used on ID cards. However, it also masks the millions of innovations that are often much smaller in scale, do not involve an invention, or are not necessarily exploited in the same commercial sense. Not included in this definition are innovations such as changing customer expectations regarding the purchase of airline tickets or dramatically improving waiting times at accident and emergency departments through improvements in patient screening. This alternative definition also has a strong technology focus because many inventions are technology based. Replacing the term *invention* with *creativity* makes the definition more applicable to organizations not actively engaged in product innovation. Therefore, a more encompassing restating of this alternative definition might be this one:

$$Innovation = Creativity + Exploitation$$

EXAMPLE: In 1923, John Logie Baird invented the television. Before its existence there had been no desire for it, but once the invention happened it established something new that never existed before. The exploitation of this invention was not instantaneous; it took decades for the TV to invade the domestic market. Broadcasting and production companies had to be set up to provide the necessary content for viewing. In hindsight, although the television did take time to exploit, it became an innovation that not only changed how we entertain ourselves but also influenced the way we live. The new flat-screen television, on the other hand, is an innovation that has been more direct in its exploitation. It meets existing customer demand for slimmer large-screen television sets. This came about because small-apartment living could not accommodate the traditional cathode ray tube (CRT) TVs. The desire for large-screen TVs led to the development of plasma and liquid crystal display televisions. The innovation has been so successful that it has resulted in a disruption in the industry, with CRT-based TVs becoming obsolete. When Philips invented the interactive TV in the 1980s, some analysts viewed it as another innovation by a company renowned for its innovation processes. They argued that it could destroy the traditional CRT television market. However, customers found the interactive TV too expensive and too cumbersome to use, and it failed to make the transition from being an invention to an innovation.

INNOVATION AND CHANGE

Although we view innovation as resulting in change, it is incorrect to equate innovation with all forms of change. In order for change to qualify as innovation, it must have some degree of desirability and intentionality (West & Farr, 1990). When we examine the output of innovation and change, another difference becomes apparent. This is that change can have a positive or negative impact on the organization, whereas innovation by definition must be positive because it must add value to the customer. Therefore, we may conclude that although all innovation can be viewed as change, not all change can be viewed as innovation.

INNOVATION AND FAILURE

One of the first writers to emphasize the importance of innovation was Schumpeter (1942), who described innovation as "creative destruction" that is essential for economic growth. Innovation is essential for helping organizations grow. Growth is often measured in terms of turnover and profit, but growth can also occur in knowledge, human experience, and the efficiency and quality of products, processes, and services. The innovation process will naturally involve unsuccessful ideas. These are seen as a natural byproduct of the innovation process. In order for some ideas to succeed, many more must fail. Organizations can learn from these failures and bring new knowledge (and sometimes technology) to use in future innovative actions that may benefit the organization. Organizations that can successfully sift out the good ideas from the bad will be more adaptable than those that cannot do so. In managing the innovation process, destroying poor ideas is often as important as nurturing good ones. Destroying poor ideas early on allows scarce resources to be released and refocused on new ideas.

EXAMPLE: Merck's product Mectizan, used to treat river blindness in developing countries, has been of huge benefit to patients: It literally allows the blind to see. However, in economic terms the product innovation has cost Merck money because the company has distributed the drug for free as part of its corporate social responsibility policy (because people with the disease usually cannot afford to pay for the drug). This example highlights the difficulty of defining innovation based on narrow metrics such as profit.

INNOVATION AND ENTREPRENEURSHIP

The terms *entrepreneurship* and *innovation* are often used interchangeably, but this is misleading. Innovation is often the basis on which an entrepreneurial business is built because of the competitive advantage it provides. On the other hand, the act of entrepreneurship is only one way of bringing an innovation to the marketplace. Technology entrepreneurs often choose to build a startup company around a technological innovation. This will provide financial and skill-based resources that will exploit the opportunity to develop and commercialize the innovation. Once the entrepreneur has established an organization, the focus shifts toward its sustainability, and the best way that this can be achieved is through organizational innovation. However, innovation can be brought to market by means other than entrepreneurial startups; it can also be exploited through established organizations and strategic alliances between organizations.

INNOVATION AND CUSTOMERS

An innovation must add value to customers to make them purchase or consume the product or service or perceive an improvement. An important part of the exploitation process is ensuring that the innovation adequately fulfills prospective customers' needs. The better the innovation fulfills customer needs, the more likely customers are to adopt it. A common mistake technology companies make is to focus on the technological capability of their offering rather than on how that technology can satisfy customer needs. It is important to emphasize that a customer is anyone who purchases or uses a product or service. Customers can include students who purchase a book in the university bookstore, patients who use services in a hospital, or members of the public who use the services of a local library. Customers can also be internal to an organization. University lecturers who offer a service to students are in turn customers of the library, for example. Doctors who deliver a service to patients are also customers of support laboratories, and librarians are customers of the library's computer service department. When we use the term *organization* in this book, we refer to the organization around which innovation is focused. This can be an entire company, a department within a company, or a team of individuals.

EXAMPLE: The development of the blockbuster drug Viagra transformed its creator, Pfizer, into one of the world's leading pharmaceutical companies. However, if an alternative use other than the original focus on heart conditions such as angina and hypertension had not been found, then the company would have written off millions of dollars of R&D investment. It is not only the innovative idea but also how an organization manages the exploitation of that idea and its fit with market need that determine success.

INNOVATION AND KNOWLEDGE

Innovation is built on a foundation of creativity and sometimes on invention, resulting in the creation of new knowledge and learning within the organization. Even when failures occur, the learning gained can be a valuable asset for the organization. The scope of innovation exists primarily within the realm of the individual and the collective knowledge of the organization. This has become increasingly evident as the complexity of technology and markets has increased. Therefore, the knowledge reservoir of the organization determines the type and level of innovation possible. If an organization's culture and routine are capable of capturing knowledge from past failures, then future innovative efforts will not repeat the mistakes of the past. Organizations that develop such knowledge systems are in a better position to store and share this knowledge so that it will improve the innovation process through enhanced idea generation, better decision making, and more effective exploitation. In this way, all ideas, whether successful or not, can contribute to the organization's long-term success.

INNOVATION AND SOCIETY

Innovation is an attribute that is beneficial to a large society such as a nation or region. Not only can innovation introduce new products and services that enrich the lives of individuals both nationally and internationally; it can also contribute significantly to economic growth. Process innovation also increases the amount of economic growth by providing cost competitiveness within a nation and attracting investment by organizations that establish bases there. National economies develop through the innovation and manufacturing abilities of their organizations and from selling the resulting innovative products on the global market. These activities not only bring increased revenue streams into the economy, increasing the gross domestic product, but also provide employment opportunities. On the other hand, innovation can have a negative impact on society by wiping out traditional industries or having other unintended side effects. For example, although a certain chemical innovation may allow farmers to grow more crops per acre, it may also pollute the environment, kill wildlife, and even cause human health problems by working its way up the food chain. In order to balance the advantages and disadvantages of particular innovations, specific regulatory bodies such as the Food and Drug Administration have been established; if side effects are deemed to be too dangerous, the product can be blocked from reaching the market. Many national governments have established agencies to promote and foster a more innovative culture in order to increase wealth and reduce costs for the nation. In this respect, performance indicators such as gross national product, export sales, direct foreign investment, R&D expenditure, employment levels, and new business startups suggest the innovative potential of a large society.

Drivers of Innovation

Various factors encourage an organization to innovate. Each of these drivers demands continuous innovation and learning so that the process can be repeated continuously. These drivers also help to create a sense of urgency around the need to create new organizational goals and generate new ideas for meeting these goals. These drivers can be summarized as follows:

- Emerging technologies
- Competitor actions
- New ideas from customers, strategic partners, and employees
- Emerging changes in the external environment

EMERGING TECHNOLOGIES

These have the potential for significant innovation across the organization and can be the basis for innovative products, processes, and services that can revolutionize the fortunes of an organization. In the past, organizations developed technologies in large R&D laboratories; however, in today's environment the sources of emerging technology are often far too prolific for any one organization to develop internally. Consequently, organizations expend more resources scanning the environment for potential technological opportunities. Sources of emerging technology can include universities, high-technology startups, and competing organizations.

COMPETITOR ACTIONS

The innovative actions of competitors and other organizations can be another driver of innovation. Competitors can provide a benchmark regarding which projects and initiatives to pursue. Copying competitor innovations reduces risk because the products may have already been adopted by the market. Although such behavior is unlikely to increase market share, it can be effective in maintaining the status quo by counteracting the advantage to the competitor.

NEW IDEAS

In the past, innovations were developed from the insights of a small number of designers and engineers. Now, however, with greater technological complexity and market segmentation, modern organizations are engaging as many stakeholders as possible in the innovation process. This can result in increased scanning capabilities and better information about market needs. Engaging employees, suppliers, customers, and other lead users can reveal new opportunities that otherwise might have gone undiscovered.

EXTERNAL ENVIRONMENT

All organizations are affected by changes in their external environment; these changes can be another driver of innovation. Environmental changes can occur because of competitor actions that have revolutionized the business environment or can happen through macro shifts in the political, economic, cultural, or technological environment. As organizations struggle to realign with their new business environment, they must innovate their products, processes, and services accordingly.

EXAMPLE: After the terrorist attacks of 2001 in the United States, governments across the globe imposed greater security requirements on the airline industry. The initial impact of this new environment was chaos at airports, long queues, and customer confusion. Airlines and airport authorities had to innovate their processes to meet these new customer requirements to remain competitive.

Categories of Innovation

The term *innovation* is often associated with products. When we think about innovation we think about a physical product: a television, car, or digital music player. However, innovation can also occur in processes that make products, services that deliver products, and services that provide intangible products. Many services don't involve physical products at all. For example, a hospital or government department offers a range of services without producing products. In this book we focus on innovation in the organizational context, and although product innovation is an important part of this, it is still only a part. We can say that innovation relates to products, processes, and services.

Product innovation is about making beneficial changes to physical products. Examples include

- Introducing a new screen size for TVs
- Changing from a CRT TV to a flat screen
- Adding functionality such as Internet access to TVs

Process innovation is about making beneficial changes to the processes that produce products or services, as for example,

- Building new systems that assemble a TV set faster and cheaper
- Redesigning the assembly line so that TVs can be manufactured more reliably
- Outsourcing the production of the plastic covers on TVs so costs can be reduced and quality improved

Service innovation is about making beneficial changes to services that customers use. Examples include

- Changing the way dealers sell new TVs in order to cut costs
- Changing the way customers get rid of their old TVs by introducing a take-back policy
- Offering credit finance options to allow customers to purchase TVs

Each of the innovations just listed adds value; if customers recognize this, they will return to purchase or use similar products or services in the future. A key characteristic of a product, process, or service is the degree of tangibility of the product and the degree of interaction with the end customer or consumer. For example, product innovation involves innovating tangible products. This is an activity in which most consumers have little involvement. The first time a customer experiences the product is when it is brought into the market and the customer purchases it. On the other hand, services typically involve intangible products such as banking and serving fast food. Unlike with products, the customer has a high degree of contact and interaction with services.

Other differences between products and services include the following:

- Inventory: Products can be stored; services cannot.
- Response time: Products have a longer lead time; services have a shorter lead time.
- Resources: Products tend to be capital-intensive (e.g., machines), whereas services tend to be labor-intensive.

Figure 1.1 illustrates these differences along the two main dimensions: the degree of tangibility and the degree of customer contact. We will now look at each of these innovation types in greater detail.

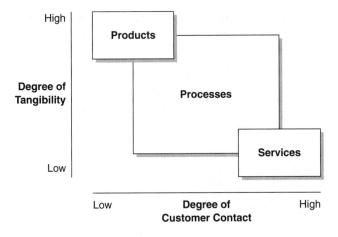

Figure 1.1 Tangibility and Customer Contact

Product Innovation

Product innovation is about making beneficial changes to physical products. Related terms that are often used interchangeably include *product design, research and development,* and *new product development* (NPD). Each of these terms offers a particular perspective on the degree of changes to products. The degree of change can include the following (Wheelwright & Clark, 1992):

- Incremental improvements
- Additions to product families
- Next-generation products
- New core products

Established organizations typically have a portfolio of products that must be incrementally improved or adjusted as problems are identified in service or as new requirements emerge. It is important that they also work on additions to the product families. One of the main activities of the product design team is the work it performs on next-generation products or new models of products. They may also work on designing radical new products or new core products that expand the portfolio significantly and often involve radically new processes to create them. These new core products ideally offer the organization the possibility of major increases in revenue and growth, which can also create the potential of a temporary monopoly in the market. The product development process for next-generation and new core products follows a familiar cycle in most organizations (Cooper, 2000):

1. Ideation

2. Preliminary investigation

3. Detailed investigation

4. Development

5. Testing and validation

6. Market launch and full production

Each of these steps involves interaction with customers, who may participate in idea generation and feature recognition. Key performance criteria in the design process revolve around the following (Smith & Reinertsen, 1995):

- Time to market

- Product cost

- Customer benefit delivery

- Development costs

These criteria can be traded off against one another. For example, development costs can be traded against time to market, customer benefits can be traded against product costs, and so on. Three design methods have established themselves as providing a management system for effective product innovation: phase review, stage gate, and product and cycle time excellence (PACE).

PHASE REVIEW

This method divides the product development life cycle into a series of distinct phases. Each phase comprises a body of work that, once completed and reviewed, is handed over to the next phase. No attention is paid to what may or may not happen in the subsequent phases, mainly because of a lack of expertise or exclusive focus on the tasks in the current phase. The phase review method is a sequential rather than concurrent product design process; that is, each phase is executed and completed before the next phase can begin. Phases typically are carried by different functions or departments within the organization. All tasks, decisions, and tradeoffs are made solely in the context of the phase being executed. A significant criticism of this approach is the poor coordination between phases, which can result in significant delays and rework.

STAGE GATE

The stage gate method is a concurrent product design process that follows a predetermined life cycle from concept generation to market launch (Cooper, 2000). The stages in this method are primarily cross-functional. Stage gates appear at the end of each stage, where a design review takes place. Each stage gate reviews the agreed deliverables for completion at the end of the stage, a checklist of the criteria agreed for each stage, and a decision about how to proceed from a particular stage. This method identifies a number of roles for people involved in the process, including gatekeepers, who are typically senior managers. Other features of the stage gate process include "fuzzy front-end" stages of customer opportunity identification, which incorporate gates that are contingent on future events.

PACE

The PACE method is concerned primarily with developing product development strategies (McGrath, 1996). The method links product strategy with the overall strategy and vision of the organization. A key feature is deploying the voice of the customer throughout the product design process. Strategies are divided into six product strategic thrusts: expansion, innovation, strategic balance, platform strategy, product line strategy, and competitive strategy.

Product innovation methods and processes are one element in an organization's mission to create value for customers. Too often functional groups within organizations have focused exclusively on the NPD process in their department as an end in itself rather than taking a broader business perspective. Interaction with functions such as marketing, warranty, manufacturing, and senior management offer design teams a more holistic perspective in the design process. This also ensures that the goals of the design process form strong relationships with the goals of the organization as a whole. Taking this broader perspective can encourage design teams to engage in new core product development, that is, to develop products that are radically different from what already exists in the organization.

As customer needs change and as markets adapt to a changing competitive environment, design teams often fail to recognize changes or disruptions to existing product requirements (Christensen, 1997). Successful organizations are capable of taking a broader perspective, recognizing the potential of disruptive technologies and then creating new products that meet the unforeseen needs of customers.

Process Innovation

Process innovation can be viewed as the introduction of a new or significantly improved method for the production or delivery of output that adds value to the organization. The term *process* refers to an interrelated set of activities designed to transform inputs into a specified output for the customer. It implies a strong emphasis on how work is done within an organization rather than what an organization does (Davenport, 1992). Processes relate to all operational activities by which value is offered to the end customer, such as the acquisition of raw materials, manufacturing, logistics, and after-sales service.

In the 1970s and 1980s process innovation gave Japanese industry a competitive advantage that allowed them to dominate some global markets with cars and electronic goods. Similarly, process innovation has allowed organizations such as Dell and Zara to gain competitive advantage by providing higher-quality products, delivered faster and more efficiently to the market than by the competition. By concentrating on the means by which they transform inputs such as raw materials into outputs such as products, organizations have gained efficiencies and have added value to their outputs. Process innovation allows some organizations to compete by having a more efficient value chain than their competitors have.

Process innovation has resulted in organizational improvements such as lower stock levels, faster, more agile manufacturing processes, and more responsive logistics. Organizations can improve the efficiency and value of their processes with a vast array of different enablers. Although the use of these enablers is contingent on the organizational context, many offer the potential for enhanced process performance. The application of technology such as robotics, enterprise resource planning systems, and sensor technologies can change the process by reducing the cost or variation of its output, improving safety, or reducing the throughput time of the process. The application of certain human resource practices can improve the quality of the process, enhance motivation, and allow increased complexity through greater flexibility. Similarly, as the raw material input to the process is altered, costs can be reduced or performance parameters improved.

A number of common approaches to process innovation have emerged through the work of operations and quality management movements over the past 20 years. Although these may not be applicable to all organizations, they can stimulate the innovation process. The more common approaches include just-in-time, total quality management, lean manufacturing, supply chain management, and enterprise resource planning.

JUST-IN-TIME

This approach originated in Japan and was originally designed to improve high-volume production by reducing setup times and other forms of inefficiency such as high inventory. Improvement is achieved through coordination of the flows of materials through a process so that the right material arrives in the correct location just as it is needed.

TOTAL QUALITY MANAGEMENT

This seeks to improve the quality of an organization's output by eliminating defects, that is, by introducing systems that prevent defects from occurring in the first place. It also engages all employees in the effort toward continuous improvement. It focuses on all aspects of organizational quality rather than just manufacturing quality and encompasses the entire organization. The approach is also characterized by the development of standards such as ISO 9000 and accreditation of organizations to these standards by accreditation bodies.

LEAN MANUFACTURING

This approach seeks to reduce all forms of waste across the total organizational system in order to increase value. It identifies seven forms of waste: transportation, inventory, motion, waiting time, overproduction, unnecessary processing, and defective products. The approach encourages organizations to identify and remove waste that reduces value within their processes by using techniques such as continuous improvement, pull systems, total quality, flexibility, waste minimization, and an integrated supply chain.

SUPPLY CHAIN MANAGEMENT

This approach focuses on managing the flow of materials and information across the entire value chain, from supplier to customer. This encourages organizations to enhance integration with suppliers and customers and establish longer-term relations. It also enhances processes by reducing overall cost and increasing value added and responsiveness to the end customer.

ENTERPRISE RESOURCE PLANNING

This approach integrates all information and processes of an organization into one holistic system. This usually relies on large software systems that facilitate the identification and planning of all necessary resources and activities across the organization in order to deliver the product or service to the customer. Such systems help identify the bottlenecks and waste within the organization's processes and support the work of other approaches such as supply chain management, just-in-time, and lean manufacturing.

Service Innovation

Service innovation is about making changes to products that cannot be touched or seen (i.e., intangible products). Services are often associated with work, play, and recreation. Examples of this type of service include banking, recreation, hospitals, government, entertainment, retail stores, and education. In the past decade a vast number of knowledge-based services have been offered through Web sites. These services involve intangible products, have a high degree of customer interaction, and are usually activated on demand by the customer. Defining a service can be somewhat problematic. Some define service as a sequence of overlapping value-creating activities. Others define service in terms of performance, where client and provider co-produce value. There are three types of services operations:

- Quasi-manufacturing (e.g., warehouses, testing labs, recycling)
- Mixed services (e.g., banks, insurance, realtors)
- Pure services (e.g., hospitals, schools, retail)

Services can clearly involve products that form an extended part of the product life cycle, from initial sales to end-of-life recycling and disposal. Service industries in areas such as finance, food, education, transportation, health, and government make up most organizations in any economy. These organizations also need to innovate continuously so they can increase levels of service to their customers.

A key attribute of a service is a very high level of interaction with the end consumer or customer. The customer is often unable to separate the service from the person delivering the service (sometimes called inhomogeneity) and so will make quality assumptions based on impressions of the service, the people delivering the service, and any product delivered as part of the service. Another characteristic of some service organizations is that their output may be perishable; therefore, the product must be consumed as soon as possible after purchase. Therefore, the timing of the delivery and customer perception of quality are crucial to success.

The concept of service quality is of particular relevance. The unique characteristics of services, such as intangibility, customer contact, inhomogeneity, and perishable production, also offer significant scope for innovation. Another major driver of service innovation comes from the possibilities afforded by new information technology platforms, particularly the Internet. The Internet is a valuable resource on which new service relationships between organizations and their customers are being developed every day.

EXAMPLE: Dell Computer Corporation uses many of the same design and manufacturing systems as their competitors, but they differ significantly in terms of how they serve their customers' needs. What originally set Dell apart from its competitors and arguably helped build its significant market share was its strategy of skipping the middleman (i.e., the sales agent) and allowing customers to configure their computers to their own requirements. This was coupled with a manufacturing system that could produce the goods very quickly and provide an all-important service and distribution process. Dell's service was also significantly boosted by use of the Internet and the availability of the Internet to potential customers.

Product and Process Innovation

Process innovation often is viewed as less important than product innovation. Whereas new product development has been the focus of much attention, process innovation has come under the headings of operations management and quality management in the literature. As highlighted by Utterback (1996), process innovation becomes of increasing importance relative to product innovation for organizations once the dominant design of the product has been established. Therefore, over the life of an industry, process innovation is of equal importance to product innovation when organizations seek competitive advantage. Process innovation can even affect product innovation when it results in improvements to the process that can inspire further product innovation. All organizations are constantly trying to develop their processes to reduce cost, improve output quality, reduce lead time, or increase value for the customer. Only certain organizations engage in product innovation, and this occurs only periodically as they develop a new product or engage in product enhancement. The greater attention paid to product innovation may result from the following:

- Individual product innovation projects are often of longer duration and greater investment than those of process innovation.

- Product innovations are more visible to the external market than process innovations.

- Product innovations are viewed as the domain of the R&D and design departments; alternatively, process innovation is viewed as the domain of the operations and quality departments.

Irrespective of the reasons behind this mindset, organizations must realize the potential offered by both product and process innovation. Every new product must be produced before it reaches the customer. If the production process cannot produce a product at the right level of cost, quality, and reliability, then the product innovation can be rendered useless. Most product innovation takes place at the early stages of the industry life cycle when numerous designs are tried and tested before the product becomes established in the product portfolio. After a certain time period, the product reaches a stage of dominant design (Utterback, 1996). After this point, the rate of product innovation decreases as mindsets are constrained by the dominant design, and the relative importance of process innovation increases across the sector as companies try to find better and more cost-effective ways to produce a marketable product. Over the life cycle of the product, the scope of process innovation decreases as the optimum configuration of production process is achieved (Figure 1.2). The end of the life cycle typically is characterized by a disruptive shift that makes existing products and processes obsolete and resets the innovation cycle back to focus on product design.

Example: Disposable baby diapers were first invented in the 1950s; they achieved only 1% market penetration because of high costs and poor performance (they leaked). The performance problem was solved after a year with more absorbent materials, but for another 10 years the product remained too costly for most consumers. Eventually a new and very complex process was developed that could produce diapers at a cost that most customers were prepared to accept. The product became a success only after the optimum process had been developed.

Radical and Incremental Innovation

The definition of innovation does not refer to the size and scope of the change to the product, process, or service. For example, introducing color television in the mid-1960s was clearly a major or radical change to the

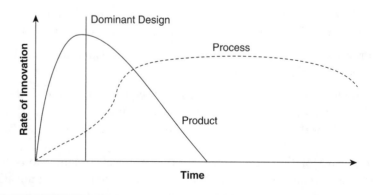

Figure 1.2   Product and Process Innovation

SOURCE: Adapted from Utterback (1996).

established black-and-white TV market. But what if a smaller change were made, such as changing the material of the television cabinet? Innovation can be classified as either radical innovation or incremental innovation.

RADICAL INNOVATION

Radical innovation is about making major changes in something established. Focus is significant in relation to this issue. A change can represent a radical innovation when viewed at a technological level, but the impact may be only incremental when viewed at an organizational level. When we examine innovation, it is the impact at this level that we are interested in. The term *radical* often refers to the level of contribution made to the efficiency or revenue of the organization (MacLaughlin, 1999). For example, by introducing the flat-screen television, manufacturers radically increased the demand for such products. We can visualize radical innovation as a step change in some measure of growth such as revenue or efficiency (Figure 1.3). Most organizations engage in some form of radical innovation over their lifetime.

Radical innovation can threaten to transform the industry itself by destroying the existing market and thus creating the next great wave (Christensen, 1997; Utterback, 1996). Undertaking radical innovation can bring dramatic benefits for an organization in terms of increased sales and extraordinary profits, but it is also highly resource intensive and risk laden. Companies in the pharmaceutical industry can invest more than $400 million in developing a new drug (Light & Lexchin, 2003) and have no guarantee that it will ever pass clinical trials and make it to the marketplace. Because of the turbulence of the external environment, it is difficult for any company to say that a potential innovation will result in a radical impact; they can only pursue the innovation with the knowledge that the scope exists for radical impact.

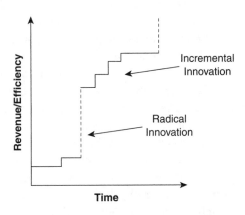

Figure 1.3   Radical and Incremental Innovation

Example: Philips invested significant resources—time and money—in the development of its interactive TV. Customers did not purchase this product in sufficient quantities to allow Philips to reach its revenue target; in other words, the new product failed to be adopted by the market. Not only did Philips lose money, but time was lost in coming up with a better innovation. If interactive TV had succeeded at that time, Philips would have had an enormous head start over its competitors and the possibility of creating a step change in its revenues.

INCREMENTAL INNOVATION

Although radical innovations often make headlines, most organizations spread the risk associated with innovation by also looking for small or incremental innovations to their products, processes, and services. In fact, some companies shy away from radical innovations altogether, preferring instead to invest in incremental innovation. Incremental innovation is less ambitious in its scope and offers less potential for returns for the organization, but consequently the associated risks are much less. Apart from using fewer resources, incremental innovations consist of smaller endeavors, making them easier to manage than their larger counterparts. Incremental innovations such as increasing television speaker power or screen size often lead to small changes in growth. However, an organization may have to undertake more and more of these types of innovation to achieve the necessary growth to survive. If an organization successfully implements enough incremental innovations, then it can sometimes lead to the similar levels of growth driven by radical innovations. The drivers of incremental innovation initiatives can include approaches to continuous improvement such as lean manufacturing, total quality management, and world-class manufacturing.

There are advantages and disadvantages to both incremental and radical innovation. Radical innovation has the advantage of creating a step change in growth. The disadvantage is the high level of risk and high cost of failure. The advantages of incremental innovation are lower risk and the possibility of achieving small degrees of growth. However, the disadvantage compared with radical innovation is the slowness to reach growth targets before competitors, leading to a loss of competitive advantage. Most organizations adopt a dual approach to the size and scope of their innovation activities. There are usually many incremental innovations going on at the same time, yielding short-term results. They may also develop some potentially radical innovations that may yield significant results in the medium to long term. Therefore, the innovative effort of an organization consists of a portfolio of innovations rather than just one specific project. As a result, decisions involving the innovation process become much more complex, as does the attitude of the organization toward risk taking in order to fund the innovations.

OTHER CLASSIFICATIONS

Defining the scope of innovation as either radical or incremental is simplistic and can be problematic. An alternative product-oriented view of innovation views its scope as consisting of four levels (Olson, Walker, & Ruekert, 1995): products new to the world, line extensions, products new to the organization but not the market, and product modifications.

Another perspective promoted by Tidd, Bessant, and Pavitt (2005) discusses the scope of innovation using the four alternative labels *discontinuous, architectural, modular,* and *incremental.*

EXAMPLE: RFID tags are now being used as a replacement for bar codes on all types of products. The current process of purchasing goods in a supermarket is well known to most of us. We wait in line at the supermarket checkout and scan the bar code of each individual item before paying. RFID tags can be detected remotely by radio receivers. If all the items in our basket have RFID tags, then all we need to do is push the shopping basket under such a receiver. It will remotely detect every item in the basket. The value for customers is shorter lines at the checkout or even the replacement of checkouts altogether. Suppliers and supermarkets that adopt RFID tags will have the potential to attract more customers than their slower-to-innovate rivals.

Disruptive Innovation

Every now and again a radical innovation is introduced that transforms business practice and rewrites the rules of engagement. In other words, business practice across an entire industrial sector changes radically. Christensen (1997) defines these types of innovations as disruptive innovations. Disruptive innovation often occurs because new sciences and technology are introduced or applied to a new market that offers the potential to exceed the existing limits of technology. Research laboratories usually are the source of disruptive technologies. Many companies watch out for the outcome of this type of technology, and from this they choose potential winners that are quickly adopted for new products and services. Some large companies, such as Intel, have their own internal research laboratories but also work in cooperation with universities and other such organizations in order to develop the latest disruptive technologies, which can take many years to develop and exploit successfully.

The driving force in creating disruptive technologies is the same as for any innovation, that is, to add value for customers that will encourage them to purchase products and services over and over from the same organization. However, determining the appropriate technology trajectory to pursue can often be difficult because there are numerous options. The task becomes even more difficult because disruptive technologies initially appear at a lower performance level than that of the existing technology (Figure 1.4). Rafii and Kampas (2002) recognize the difficulty for management of trying to distinguish signal from noise with respect to future disruptive threats. They suggest that organizations should harness their collective wisdom through a structured decision-making method to filter out the noise and identify potential disruptive threats that will inflict damage. Through their formalized decision-making approach, they believe that an organization's process of scanning for disruptive threats can become more rigorous and can increase the chances of formulating an appropriate organizational response.

Disruptive innovations can often appear in niche markets and can take time to show their real potential in dominating mainstream markets (Christensen, 1997). Organizations must take extra care to select the correct disruptive innovations to pursue. Pursuing the wrong technology can waste scarce resources and place the organization in a position of significant competitive disadvantage. Interestingly, organizations that are industry leaders of the preceding technology generation often fail to make the transition to the new industrial reality (Utterback, 1996). One can put this down to the mistake of pursuing the incorrect technology; however, one well-known commentator promotes the view that these dominant companies are overly supportive of their own business model and do not see the need to embrace the new reality (Christensen, 1997). When discussing breakthrough innovation, O'Reilly and Tushman (2004) highlight the fact that established organizations often struggle to achieve a successful balance between developing radical and disruptive innovations while still protecting their traditional business operations. They emphasize that although this dual focus is difficult, the creation of a structure of

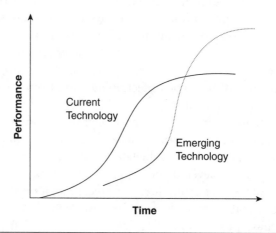

Figure 1.4   The S-Curve for Performance

organizationally distinct units that are tightly integrated at the senior management level can facilitate pursuit of this dual focus. They call this type of structure an ambidextrous organization because it provides an effective framework for organizations to pursue pioneering innovation while still achieving incremental gains in their traditional business. The ambidextrous organization is discussed further in Part IV.

There are many examples of disruptive technologies introduced in recent years, including the following:

- Data storage disks
- Digital photography
- RFID tags
- Digital media (music and video)
- The Internet and the World Wide Web
- Text messaging and the mobile phone

Arguably the largest disruptive technology to emerge has been the World Wide Web. The Web has disrupted products, processes, and services across most industrial sectors. Products such as televisions can now be Web enabled, offering customers a host of new services such as video downloads and Internet browsing. Processes such as the manufacturing of televisions regularly use the Web to source materials, receive orders from television dealers, and track sales in real time. Services such as movie rentals now use the Web to offer customers the latest movies, which can be downloaded directly to a television or computer on demand. Thus, organizations need to be aware of the threats posed by disruptive innovations in their industry and be prepared to react quickly if necessary.

EXAMPLE: The movie rental business has existed for the past 30 years or so. First, the industry had to decide which dominant design to choose: Betamax or VHS. When VHS eventually won out, organizations that had adopted the Betamax standard found their expertise obsolete and ended up lagging significantly behind the industry leaders. The next shift occurred in the early 1990s, when the DVD won out over other alternatives such as the compact video disk. Today the industry again finds itself at a crossroads: Which technology to adopt? And what will be its impact on growth? Options include broadband downloads to a personal computer, universal media disks, Blu-Ray, and other types of storage media. Like the majority of other discontinuous shifts, all these technologies are currently inferior to the DVD in certain ways, yet they address certain unfulfilled customer needs such as the nuisance of having to go out to rent a movie and remembering to return it. Where the industry will go in the future is still uncertain; the only certain thing is that in 5 years the DVD as we currently know it will be as obsolete as the VHS tape.

Innovation and Operations

Simplified to its most basic level, an organization can be said to consist of two core activities: operations and innovation. Operations are all the activities that provide an existing service or product to a customer, including manufacturing, human resources, and material planning. Operations usually form the mainstream activities of any organization and are focused on the here-and-now needs facing the organization. Innovation, on the other hand, consists of all activities that change operations and are focused on the future needs that the organization will face. Activities such as product design, process engineering, and system analysis are processes of innovation. In many organizations, tension exists between operating the system that provides existing products and services for the customer and changing the system in order to add more value to the customer, which disrupts the established operations. Some organizations often try to maintain the status quo and resist innovative change. Innovative organizations, on the other hand, embrace the challenge of maintaining a balance between serving the needs of the existing customer and meeting the future needs of the market.

Metaphorically, operations can be seen as a rotating wheel. If no changes are made to products, processes, or services, then the wheel will continue to turn; the organization will continue to serve customers with existing products and services for a time. But changes will occur outside the organization. Customer expectations, business conditions, and competitors will change. Innovation is about oiling the wheel, making it run more smoothly and more efficiently—making enhancements to products, processes, and services that better meet the changing needs of customers.

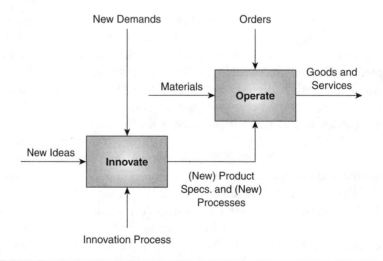

Figure 1.5 Operations and Innovation

A useful way to visualize the relationship between the operations and innovation activities is presented in Figure 1.5. This figure illustrates an activity diagram with two activities labeled with active verbs: *Operate* and *Innovate*. The first activity, Operate, converts orders from customers into goods and services that fulfill customer orders. Other inputs include raw materials and resources such as product specifications and processes. All three inputs are needed to produce the output. The second activity, Innovate, is about making beneficial changes to products, processes, and services. It has an entirely different set of inputs and outputs. The major input is new demands from customers and other major stakeholders. These new demands can stimulate the generation of ideas, and new ideas can be converted into new product specifications and new processes for the Operate activity. The relationship between Innovate and Operate is symbiotic. One serves the customers; the other changes the way the customer is served. This figure is simplified and incomplete, but it illustrates the two core activities in any organization. Although the majority of innovation output is focused on improving operations, certain innovation output is focused on the innovation activity itself. Thus, the organization not only strives to change operations in terms of new products, processes, and services but also changes the means by which the innovation itself is achieved. This book is focused primarily on the innovation management process and how it can be enhanced in organizations so that it can be more effective and efficient.

EXAMPLE: Innovation is applicable to any type of organization, profit or nonprofit, and also at various levels within the organization, from management teams and departments down to large project teams and even individuals. Innovation can occur in products, processes, and services and can involve radical or incremental changes that help an organization to grow in many ways. Table 1.1 presents a list of organizations that have developed their own innovation plans. The names of the organizations have been altered to protect confidentiality, but the URLs of similar organizations are real. Notice that some organizations are entire businesses, whereas others are departments within larger organizations.

Summary

Every organization needs to innovate. Innovation is as relevant to a hospital, movie theater, or press office as it is to a manufacturing plant or product design department. Innovation is the process of making changes to something established by introducing something new that adds value to the customer. Innovation is not to be confused with invention. It can be radical or incremental, and it can apply to products, processes, and services.

Table 1.1   Innovation Plan Titles

Organization	Plan Title	Benchmark
Reflective Display Films	Innovation Plan 2003–2006	www.reflexite.com
Galway Tables	Development Plan 2009	www.blueberriespine.co.uk
Beglin Entertainment Centre	Play Development Plan 2006–2009	www.citylimits.ie
Connemara Distillery	Change Program 2006–2013	www.bushmills.com
Qualtrans Translations	Strategic Plan 2010	www.translations.com
Car Consultants	Five-Year Development Plan	www.autofindersusa.com
Dynocorp Nightclub Group	Innovation 2006–2010	www.vcgh.com
HarPer Sculpting	Innovation 2000	www.gymamerica.com
Medical Device Manufacturing, Computer Services Department	Continuous Improvement Plan	www.medtronic.com
Precision Engineering, Engineering Department	Lean Manufacturing Plan 2k10	www.bellurgan.com
Ratio Technology, Design Department	Design Plans 2006–2009	www.raleigh.com
Moneypits Bank, High Street Branch	Branch Development	www.bankofamerica.com
Children's Hospital	Development Plan 2010	www.olhsc.ie
Ambec Resorts	Hotel Development Five-Year Plan	www.bestwestern.com

There is a special relationship between process and product innovation. The term *disruptive innovation* includes radical innovations within organizations that disrupt the way business is normally conducted. In the next chapter we will examine the innovation process and how organizations can enhance their innovative capability by managing it properly. We examine some of the change techniques that can support the application of innovation and also the impact organizational culture may have on nurturing innovative capability. The chapter also looks at innovation investment and some common reasons why innovations fail.

Activities

Applying innovation is a complex and difficult task. The activities in each chapter of this book are designed to allow you to develop your own innovation plan for a fictional organization that you will define and to struggle with some of the same decisions that any innovation team encounters when managing the innovation process. These activities can be undertaken on an individual basis, but we recommend a group approach. The output of all the activities throughout the book combines to create a comprehensive innovation plan for your chosen organization, that is, your very own case. For the purpose of these activities your organization may be based on a real organization, or it may be constructed from your imagination and based on your own experience.

To help you with the many decisions you will have to make, there is a sample plan in the Appendix for you to study. SwitchIt Manufacturing is a partially completed plan that contains tasks for you to complete. You are strongly encouraged to consider completing the SwitchIt Manufacturing case before beginning your chapter activities.

In this first activity you are required to create some simple pieces of information for your organization. The organization you create can be a large organization or a department within an organization. Examples of large organizations include hospitals, manufacturing plants, cinemas, software design houses, and local government. Examples of departments include computer services, engineering, quality assurance, human resources, and logistics. The organization you establish will need to be large in size. We recommend beginning with approximately sixteen individuals made up of managers and employees. These people will participate in developing and implementing the innovation plan you will develop for your organization. As you progress through the activities in each chapter, you can revisit this activity and add new employees and functions as needed.

Search online for a real organization that is similar to the organization you are considering developing. Make a note of its homepage address. Research this real organization's innovation effort to discover the goals it pursues, the products and services it has developed for the market, who its competitors are, and what aspects give it its competitive advantage. Choose a fictitious name for your organization, such as Fast Fasteners Ltd. or Kinder Childrens Hospital. Next, choose a name for your innovation plan that also includes a planning period (typically 1–5 years), such as Fast Fasteners Innovation Plan (2007–2010). Copy Table 1.2 into a spreadsheet and complete the fields.

Table 1.2   Create an Organization

Organization	
Name:	
Plan Title:	
Benchmark:	http://

Products and Services

Name: Name of your team (e.g., ABC Corp. or ABC Corp. Quality Dept.)

Plan Title: e.g., Innovation Plan 2007–2010 or Development Portfolio 2010

Benchmark: Web site of a similar but real organization

List the main products or services offered.

STRETCH: Other elements of this activity may include visiting a real organization that you are interested in and interviewing senior managers and team leaders regarding their goals, actions, teams, and results. Revisit this organization from time to time as you complete other elements of your innovation plan.

REFLECTIONS

- Define innovation and explain the difference between it and invention.
- Give one example of each of the following types of innovation: product, process, and service.
- Give one example of a radical innovation and an incremental innovation.
- What is a disruptive technology?
- Explain how product and process innovation are related.
- Explain the relationship between innovation and operations.

Managing Innovation 2

Innovation is about making beneficial changes, large and small, to products, processes, and services. Changes are unique to each organization. However, there are common traits in the way innovation can be managed in every organization. We explore some of these common traits in this chapter. We begin by looking at the nature of innovation and in particular a number of common techniques for innovation management. We then look at some of the traits that make some organizations excellent. Innovation culture is about the norms and attitudes that make innovation effective. This is explored in the context of looking at the positive traits of innovation culture and examining the barriers that hold it back. We then look at a number of models of innovation. The chapter concludes by introducing the concept of innovation management and how organizations can put routines in place that make innovation more effective.

LEARNING TARGETS

When you have completed this chapter you will be able to

- Explain the relationship between change and innovation
- Understand one change management method
- List some of the traits that make some organizations excellent
- Explore how culture affects innovative capability
- Explain how the culture of organizations can be adapted
- Discuss a number of models of innovation
- Discuss some of the issues around the management of innovation

Techniques of Change

Innovation results in making changes to the organization's products, processes, and services. Although not all organizational changes result in innovation, there is overlap between innovation management and change management. Consequently, certain aspects of change theory are applicable to the management of innovation. Organizational change is the process of converting an organization from its current state to some future desired state. A program or plan is created that identifies what needs to be changed, by whom, and when. The present state of the organization is usually articulated in terms of understanding current products, processes, organization structures, and so on (Figure 2.1). The future state of the organization may be illustrated through the goals, ideas, and projects that must be implemented to create change.

Over the years a number of techniques have emerged for helping organizations manage change. Some of these, such as organizational design and project management, have attempted to help the whole organization, whereas others have focused on one particular aspect of change, such as total quality management and Six Sigma. Figure 2.2 illustrates a small sample of the techniques that have emerged for managing various aspects of change in organizations. These change techniques may contribute to innovation by acting as a stimulus to employee creativity by defining their innovative ideas, projects, and goals. Similarly, the associated implementation methods of these techniques can provide valuable insights into the methods of realizing innovation actions.

Many of the techniques have emerged from particular fields or disciplines. For example, in the management field, techniques such as strategic

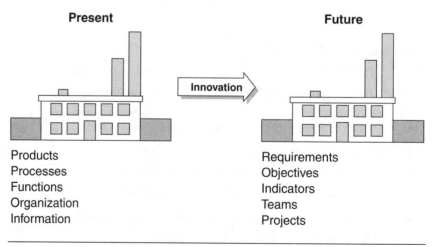

Figure 2.1   Present and Future Organizations

Figure 2.2   Change Management Techniques

SOURCE: Dooley and O'Sullivan (2001).

planning, performance measurement, and technology management have evolved. In the field of systems theory, techniques such as structured analyses and business process reengineering have emerged. All of these techniques have one thing in common: They help organizations manage and create change. Some analysts argue that as the various techniques become more familiar across the various disciplines, they begin to coalesce and converge toward a common set of techniques for change. Applying innovation, as presented in this book, is one such composite technique that builds on the following techniques. These and other techniques associated with the area of change management can contribute to the organization's ability to manage innovation:

- Strategic planning
- Performance measurement
- Creativity management
- Project management
- Knowledge management

STRATEGIC PLANNING

This technique has been around for many decades and involves organizations setting out the broad areas for change over a planning period. These areas for change are called strategies or objectives and are aimed at guiding individuals in the development of ideas and projects that achieve the objectives.

PERFORMANCE MEASUREMENT

This technique involves putting measures or indicators on certain critical aspects of an organization's performance. Performance indicators must be easily measurable. Indicators are typically used to encourage and foster change in the organization (i.e., if a particular indicator is showing poor performance, then action must be taken to remedy the situation).

CREATIVITY MANAGEMENT

This technique is often focused on the area of generating ideas and solving problems. As we shall see later, generating ideas rarely occurs in isolation. The organization typically scans the environment for opportunities and aligns the ideas it pursues with its defined goals for the future.

PROJECT MANAGEMENT

This technique focuses on the need to manage various change initiatives and tasks effectively. It involves a disciplined approach to dividing projects into tasks and scheduling these tasks according to resource availability. An important discipline within this technique is portfolio management, by which groups of projects are managed relative to one another. As with innovation management, this technique is heavily dependent on an understanding of the goals of the organization.

KNOWLEDGE MANAGEMENT

This technique focuses on how to effectively manage change by managing the information associated with change. For example, in this book innovation will be put into practice using a simple knowledge management system for managing all the key information involved in the innovation process. This knowledge management system will be implemented using tables.

Change Methods

A broad variety of methods can be used to facilitate the change process. Methods are step-by-step approaches to achieving an end goal. One method of note has been promoted by John Kotter (1996). His method outlines eight steps:

1. Establishing a sense of urgency

2. Forming a powerful guiding coalition

3. Creating a vision

4. Communicating the vision

5. Removing obstacles for acting on the vision

6. Planning for and creating short-term wins

7. Consolidating improvements

8. Institutionalizing new approaches

Establishing urgency involves looking hard at the organization's competitive position, communicating this information broadly and dramatically, motivating staff and employees, looking for leaders and champions of change, and discussing unpleasant facts openly. Forming a coalition involves developing a strong bond of loyalty between managers and the company. Creating a vision involves developing a mental image of a possible and desirable future state that is realistic, credible, and attractive and that most people can buy into. Communicating the vision involves winning the hearts and minds of individuals. All existing communication channels should be used, and clearly behavior needs to match words. Removing obstacles includes identifying resistance to change from individuals as early as possible. Creating and achieving short-term wins is necessary for boosting morale and convincing everyone that overall success is possible. Consolidating improvements involves making sure that change sticks and that things don't return to the old ways of doing things. Finally, institutionalizing new approaches involves making sure that the entire change program leads to lasting change in the organization.

Excellent Organizations

Many organizations struggle to innovate effectively, and their failures result in reduced revenues and losses of market share, customer confidence, and further investment. However, certain organizations have developed a reputation for excellence and time after time manage to grab the headlines as successful innovators. These elite organizations include companies such as Google, Intel, 3M, Black & Decker, and Apple, and each has demonstrated a consistent ability to bring innovative concepts to the market. Undoubtedly there are many smaller organizations that are also excellent and manage to create exceptional innovation in often more competitive environments. Peters and Waterman (1988) call all these organizations "excellent," and the mantle is deserved for the inspiration they give to others alone. Although these organizations originate in different sectors, excellent organizations have a number of common attributes. In their best-selling book *In Search of Excellence,* Peters and Waterman define excellent organizations as those that continuously revamp, adjust, transform, adapt, and respond to shifting customer needs, the performance of competitors, international trade realignments, and government regulation. They note eight common attributes of excellent companies. Although Peters and Waterman list mainly commercial businesses, all these traits are equally applicable to nonprofit service organizations. Many of these traits are also equally applicable to small organizations (e.g., departments, workgroups) within larger organizations. The common attributes are as follows.

A BIAS FOR ACTION

Excellent organizations experiment with new ideas. They test out ideas on customers, often within weeks with inexpensive prototypes. They have not allowed their previous successes and size stifle their ability to respond quickly to adding value to customers.

CLOSE TO THE CUSTOMER

Excellent organizations interact continuously with their customers in order to solve existing problems and generate new ideas for products and services. They listen intensely and regularly to customer needs, particularly the needs of lead users (i.e., customers whose needs and use of a particular product are exceptional).

AUTONOMY AND ENTREPRENEURSHIP

Excellent organizations support risk taking by individuals and teams within the organization and support good ideas through the many difficult stages of their life cycle. The entrepreneurship principle is extended to individuals willing to make exceptional effort and take risks to bring an idea to fruition.

PRODUCTIVITY THROUGH PEOPLE

Excellent organizations treat all individuals as potential sources of new ideas that will add value to customers and meet the needs of stakeholders. Individuals are encouraged to engage in problem solving and idea generation and execution.

HANDS ON, VALUE DRIVEN

Managers, leaders, and many individuals are hands-on types; they understand all the tasks needed to add value to customers and often spend time with colleagues in delivering services. They are value driven, striving to find more and more ways to add value to customers. Management is not a desk job that demands endless meetings and reports.

STICK TO CORE BUSINESS

Excellent organizations learn to develop skills and expertise around a core product, process, or service and rarely deviate from that business. They develop skill sets, corporate knowledge, and a culture that remains focused on a core business. Introducing a new product or process outside core business is assessed carefully.

SIMPLE FORM, LEAN STAFF

The underlying processes in excellent companies are elegant and simple. Top-level staff is lean; it is not uncommon for a corporate staff of less than 100 to run a multibillion-dollar enterprise, and many excellent organizations abandon a formal functional structure for flexible and less formal structures such as matrix or network structures.

SIMULTANEOUS LOOSE AND TIGHT PROPERTIES

Excellent companies are both centralized and decentralized. They push autonomy and discretion down the hierarchy to all individuals and teams. On the other hand, they are fanatical centralists around the few core values they hold dear and around core goals for the organization.

Peters and Waterman strongly support the overarching focus of staying close to and understanding the needs of customers and the importance of bringing individuals together through appropriate organizational structures. Individuals form a community with a common purpose that provides the ingredients for an effective innovation culture in organizations.

Innovation Culture

Culture can be defined as the pattern of shared basic assumptions that an organization has learned to use in dealing with internal and external changes

(Schein, 2004). Culture reflects the fundamental beliefs of the collective, and the organization can harness this potential as a powerful enabler of innovation. Organizations that develop a culture that is aligned with the requirements of innovation can nurture innovative actions and increase the innovative capability of the organization. However, organizations that have a culture that has a poor alignment can stifle and impede innovations, reducing the likelihood of success, and will need to work on changing their culture to a more supportive type. In striving to become innovative, organizations need to look within themselves and assess what takes the organization forward and what holds it back. Four key factors are found to either stimulate or depress the innovation activity (King & Anderson, 1995): people, structure, environment, and culture.

PEOPLE

Individuals and teams are a key resource in idea generation and implementation and ultimately in making sure that the customer's needs are met and exceeded. Leaders are responsible for leading the task of identifying requirements, setting goals for the organization, and ultimately making decisions that ensure that good ideas are executed efficiently. The structure of teams can vary, but more important is the behavior of individuals within the team. Effective teams have effective leaders, autonomy and discretion, appropriate skill sets, and a commonly understood set of objectives.

STRUCTURE

Organizational structures can be mechanistic or organic. Mechanistic structures have deep and narrow hierarchies between management and staff, responsibilities are well defined and rigid, communication is principally through the formal hierarchies, and power and authority typically are based on seniority. Organic structures have flat and group-based hierarchies, communications occur both through hierarchies and laterally across functions, employees have flexible job descriptions, and power and authority are based on ability. Large innovative organizations have a tendency toward organic structure to support their innovative efforts.

ENVIRONMENT

The environment that the organization inhabits will also influence the organization's innovation activity. The level of competition in the external environment can force organizations to innovate in order to survive. If a monopoly exists, there is little pressure on the organization to innovate.

In environments where strong links between organizations exist, synergies can emerge that result in innovations. Similarly, industry maturity and structure influence the organization's innovative activity. By enhancing its links with its external environment, the organization can identify more opportunities that may lead to innovation.

CULTURE

Although organizational cultures develop from the unique experiences and practices of an organization's members, certain routines and practices can enhance the organization's innovation activity.

Types of Culture

Handy (1985) has developed the theory of organizational cultures around four types of culture: role, power, task, and person.

ROLE

Role culture is a classic bureaucratic model dominated by formal rules, regulations, and procedures that try to ensure that everyone knows who is responsible for what. People rarely think outside their responsibilities or interfere with the responsibilities of others. Rules and regulations are regularly used to stop people from moving beyond their responsibilities. Role cultures are not generally effective for the promotion of innovation, which often requires that people take risks that include thinking outside their current roles and challenging others' responsibilities.

POWER

Power cultures are generally found in organizations that have developed around one person. They are typically found in small to medium-sized enterprises. Unlike in a role culture, decisions may be made on an ad hoc basis by one or two strong people. Limited authority and responsibility are shared with other people in the organization. All key decisions are made by the power figures. When the organization is small and staff generally share the views of the power figures, such cultures can contribute to effective innovation. This can change as the organization grows or the decisions made by power figures begin to lack the necessary impact because of major technological and market changes.

TASK

Task cultures are generally associated with matrix structures in large organizations. People report to different managers depending on the task being executed. Matrix cultures are flexible and adaptable, with high value placed on individual and group performance. Task cultures are effective for innovation in established and growing organizations.

PERSON

Person cultures appear in organizations comprising a number of highly skilled people. Such organizations are often highly decentralized and contain informal structures. Examples include architects and general practitioners. Person cultures can generate a high level of innovation for the people involved, but this may not develop into innovation for the organization as a whole because of the need to persuade colleagues and reach consensus at the various stages of development. There is a strong dependence on interpersonal relationships, and if problems occur in these relationships they can have a devastating impact on the whole organization.

Although the precise mix and manifestation of these types of culture can differ with the context, certain organizational traits provide a blueprint for configuring a suitable innovative environment. These traits influence the organizational culture and how it nurtures innovation and include leadership, agile structure, mix of individuals, adequate resources, outward focus, supporting systems, and a learning focus.

Leaders are needed who can communicate the importance of innovation to other people and remove barriers that hinder innovation. Structures are needed that allow cross-boundary collaboration that will support innovative actions through teams that possess broader skills and are less myopic in their perspective. The correct mix of individuals provides the skills and background for collaboration and creativity to occur. Organizations need to train employees with the necessary skills to enable them to be empowered and actively engaged in the innovation process. Organizations need adequate resources (both financial and human) to allow them to engage in innovation activities. Organizations also need to adopt an outward focus for their innovation activities because increased scanning of the external environment opens the organization up to new knowledge, opportunities, and insights. Supporting systems reflect the operation of the innovation process and make the process tangible. Supporting systems can support innovation activity through appropriate training in problem-solving techniques or routines for screening potential innovations. Finally, organizations need a learning focus so that learning can result in the generation of new knowledge and concepts that can be exploited in the future. The promotion of these traits within an organization supports the innovation activity and enhances the organization's innovative capability.

Barriers to Innovation

Another useful approach to understanding the culture necessary for effective innovation is to look at the barriers to innovation. Kelley and Littman (2001) identify what they call barriers and bridges to innovation (Table 2.1).

Table 2.1   Barriers and Bridges to Innovation

Barriers	Bridges
Hierarchy	Merit based
Bureaucracy	Autonomy
Anonymity	Familiarity
Clean	Messy
Experts	Tinkerers

SOURCE: Kelley and Littman (2001).

Organizations can find excessive hierarchy a barrier to innovation when decisions and ideas need to follow vertical paths through the decision-making hierarchy. On the other hand, flat organizations that are willing to accept ideas from all sources based on merit are supportive of innovative activity. Excessive bureaucracy can stifle innovation and slow the necessary innovation that sustains growth. On the other hand, greater autonomy and discretion allow individuals and teams to take the risks necessary for change. Anonymity becomes a barrier when individuals are unprepared to notice change or simply lie low when difficult decisions must be made. On the other hand, innovative organizations foster a culture of familiarity between individuals in which goals, problems, and ideas can be discussed openly. Another barrier to innovation can be creating a clean environment in which procedures are strict and everyone fulfills a particular role. Innovative organizations, on the other hand, tend to be messy and based on bringing unfamiliar things together to nurture creativity. Expertise can inadvertently stifle ideas through excessive criticism of the ideas of people who are not seen as having the appropriate skills to generate ideas. On the other hand, in an organization of tinkerers, it is common to find people who are willing to experiment, take risks, and explore new avenues and spaces.

Sheth and Ram (1987) outline additional barriers to innovation: expertise barriers, operation barriers, resource barriers, regulation barriers, and market access barriers.

EXPERTISE BARRIERS

These are created when organizations move outside their areas of expertise and specialization. Organizations and their inherited cultures are notoriously rigid when it comes to established core expertise and cannot perform at the same levels of expertise outside their main activities. Expertise barriers can be overcome through acquisitions, mergers, and research alliances with organizations that do have a proven track record in a new endeavor.

OPERATION BARRIERS

These are another innovation inhibitor. All new products must be produced. Existing operation environments that include manufacturing, sales, and service are often incapable of producing a radically new product. For example, an organization that has built a strong operational capability around metal products may find it difficult to produce plastic products. An organization that has developed strong sales and service operations around selling products to end consumers may find it difficult to provide products to other organizations. Operation barriers can be overcome by establishing new operation organizations separate from existing ones or the wholesale restructuring of the management organization.

RESOURCE BARRIERS

These exist where insufficient funds are available for innovation. Innovative organizations invest significantly in innovation and use government sources of funding as much as possible. Innovation often entails long-term risks, and these long-term risks can be very costly.

REGULATION BARRIERS

These are created by industry organizations and governments, which typically want to regulate certain processes in order to improve health, safety, and consumer confidence. Regulations tend to preserve the status quo. Regulations can be overcome through lobbying of government agencies and participation in regulatory bodies.

MARKET ACCESS BARRIERS

These barriers exist where organizations cannot access potential customers because of either physical distances or regulations such as trade barriers. These barriers can be overcome through agencies, strategic alliances, mergers, and acquisitions.

Adapting Culture

Culture is not something that can be altered quickly. An organization that tries to change its culture encounters resistance because it is ingrained in the norms and practices of employees. However, culture is a learned belief system that can evolve slowly in response to certain stimuli. Therefore, culture can be changed to better align an organization with the needs of greater innovation. One model that is supportive of cultural change is Johnson and Scholes's (2002) cultural web. This model identifies seven key elements that provide stimuli for affecting change in an organizational culture: paradigms, power structure, organizational structure, control system, symbols, stories, and rituals and routines. An organization that wants to change its culture can do so by adapting each of these key aspects and ensuring that all communicate a consistent message to employees.

PARADIGMS

Paradigms rest at the center of the organization's cultural web and are the set of assumptions commonly held to be true and accepted across the organization. Employees rely on these paradigms to undertake their day-to-day operations in a manner satisfactory to the organization. Paradigms provide stability within an organization over time and provide a frame of reference through which employees can interpret turbulence in their environment. However, inappropriate or outdated paradigms can result in employees having an imperfect perspective on their environment. Therefore, management must understand the paradigms present in their organizational collective and assess whether these paradigms are appropriate to nurturing innovation.

POWER STRUCTURE

This defines how power is distributed across the organization and the influence this power can exert. Often the most powerful group in the organization consists of those most closely aligned with the beliefs and assumptions that are held as important by the organization. These people have the ability to influence the organizational paradigms because they have the power structure necessary and are often recognized as the gatekeepers of certain paradigms. Although hierarchy is often the basis of the power structure of an organization, power can also be based on knowledge and expertise, the ability to reward and punish, or even an individual's charisma. Structuring power based on certain parameters such as knowledge and expertise can contribute to innovation and place greater emphasis on the intellectual capital of the organization.

ORGANIZATIONAL STRUCTURE

The type of organizational structure influences the behavior of people in the organization. These structures reflect the formal and informal ways in which power is delineated and highlight important relationships and communication channels. Structures that are mechanistic can lead to functional silos and poor communication, both of which can inhibit collaboration and creativity. The presence of more organic structures can encourage interaction across boundaries and influence the power structures and paradigms evident in the organization. The type of structure evident in the organization can influence the behavior and attitudes of employees.

CONTROL SYSTEM

The control system reflects where the organization places its emphasis and what it deems important. The various measures the organization pursues and the mechanisms it uses to motivate and punish certain actions are strong predictors of human behavior. Overly rigid control systems can prevent innovation by taking creativity and risk out of the organization and communicating inappropriate messages to employees. Where systems are tightly controlled, employees do not perceive the value of engaging in experimentation and discovery. Instead they choose to operate within the existing practices and maintain the status quo.

SYMBOLS AND STORIES

Culture is synonymous with the tradition and history of the organization. Stories and symbols are important mechanisms for embedding the present in the organization's history and reinforcing the beliefs and paradigms of the organization. Symbols can be powerful representations of the nature of the organization and can influence the behavior of employees. Organizations can effect cultural change by promoting stories and symbols that reinforce the desired culture and creating champions and routines that support the change. Over time people begin to relate to the new stories and symbols, which lead to shifts in beliefs and assumptions.

RITUALS AND ROUTINES

The routines and rituals practiced by the organization are often a practical manifestation of the organizational culture. These routines represent the normal mode in which the organization operates and reinforce the cultural beliefs at an operational level. The organization can influence its culture by modifying these routines and realigning the other key elements listed in this section to support the new routines. Key stakeholders will need to be convinced of the validity of the proposed change if resistance is to be avoided and the new routine accepted.

These elements are highly interrelated, and changes must be implemented across all elements if an organization is to manage cultural change. Where one element remains unchanged, it provides a reminder of the old culture and highlights the shift between the old and new culture. Change must be undertaken in an environment of openness and trust between all concerned. The beliefs of the old culture were adopted because they were proven to work, and if people feel threatened then they will revert to the old culture and undermine the new one. Only through careful management of all elements of the cultural web can an organization change its culture to one more supportive of innovation.

Models of Innovation

Innovation should be viewed as a process rather than an event. For many organizations, innovation occurs by chance, mystically appearing from random sources and being captured by the organization to generate value. Focusing on the event rather than the process makes innovation difficult to manage because every event is novel and unique. Organizations that focus on the process by which innovation occurs have been able to develop routines and practices that influence the process and its eventual outputs (Tidd et al., 2005). Although an organization can never guarantee success, it can implement practices and routines that improve the efficiency of the overall innovation process and subsequently its success. The current understanding of the innovation process has evolved over time. Rothwell (1992) defines five generations of the innovation process that have been steadily increasing in efficiency over time. These models have evolved from simple linear models to integrated and networked models. The five generations of innovation models are presented in Table 2.2.

Table 2.2   Five Generations of Innovation Models

Generation	Description	Emergence
First	Technology push model	Early 1950s–mid-1960s
Second	Market pull model	Mid-1960s–early 1970s
Third	Coupling model	Early 1970s–early 1980s
Fourth	Integrated, parallel model	Early 1980s–mid-1990s
Fifth	Integrated, networked model	Late 1990s

SOURCE: Adapted from Rothwell (1994).

FIRST GENERATION

The first generation of innovation models emerged during the economic explosion after World War II. Industry activity from this era was based primarily on the exploitation of new technological opportunities that emerged from well-resourced R&D laboratories. Examples included semiconductors and synthetic materials such as plastics. After years of rationing, the consumer was favorably disposed toward scientific discovery and technological advancement and readily embraced technology-based products that flowed into the marketplace. Government also encouraged this technology focus through funding and military contracts. The first generation of innovation model that emerged was linear, driven by significant R&D capability that pushed technologically superior products into the marketplace. Consequently, organizations developed strong competences within one or many technological platforms and focused on developing strong competence at the discovery stage of the innovation process to produce a steady stream of breakthrough technologies. Under this perspective, the dominant power rested in R&D, leading to the stereotype of the innovative company as one populated by white-coated scientists.

SECOND GENERATION

The second generation of models evolved during the mid-1960s as the postwar economic boom began to stabilize and competition for market share began to increase. As a consequence, customers became more discerning as to how competing products could best fulfill their specific needs. Consequently, marketing as a means of discerning customer needs became of increasing strategic importance for the organization and resulted in the innovation process shifting from a supply to a demand focus. Again, the innovation process was perceived as linear but was driven by the needs of the marketplace rather than technology. The implication of this second-generation model was that organizations expended resources through their marketing departments to better understand what the customer wanted and then use this need as a mechanism to pull appropriate innovations through the process. The eventual output of the pull process was better alignment with specific market needs. The power also shifted from the R&D function to the marketing function as technology was demoted to an enabler role.

THIRD GENERATION

The third generation of models emerged in the early 1970s, driven by economic stagnation and increased competition as a result of the oil crisis. Because of these increased pressures, organizations began to adopt a greater cost–benefit perspective, which filtered through to the innovation process. Another factor in the evolution of this new model was an inherent weakness in the preceding two perspectives. Organizations pursuing a technology push model produced radical innovations based on technological breakthroughs but often had difficulty finding a market use for their innovation. Similarly, organizations that pursued a market pull model had a ready market to accept their innovations, but their developments tended to be incremental and easily copied and surpassed. Consequently, the third generation of models was one of coupling, which promoted the ideal that technology push and market pull were opposite ends of a spectrum, and reality required significant interaction and tradeoffs between the two extremes. Organizations began to view their innovation process as sequential, functionally specific phases that are highly interdependent. The simplicity of a model driven by a single stimulus was abandoned in favor of a more interdependent and complex process. Organizations pursuing this coupling model endeavored to balance the competing pressures of technology push and market pull in order to achieve the optimum balance. The issue of dominant power also shifted, with power moving up a level of management in the organization because the primary focus of this generation was coordination and optimization across the various phases of the innovation process.

FOURTH GENERATION

The fourth generation of models evolved during the economic resurgence of the mid-1980s and was strongly influenced by the practices of Japanese industry that were in ascendancy at the time. Organizations were obsessed with time to market and waste avoidance because they seemed to explain the Japanese dominance. These influenced the way organizations managed their innovation process and resulted in a model that focused on integration and parallel development. Organizations also strove for better understanding of the functioning of the process to include the effect of elements previously deemed random variation and thus uncontrollable. This new model also broadened the scope of innovation to encompass the power of the supply chain, including suppliers and other organizations not traditionally included in the innovation process. The idea of a sequential process was replaced by a perspective of integrated activities that occur simultaneously and influence each other's development. This model promoted the idea of overlap between the phases, where effective knowledge exchange across boundaries was essential to success. The management of innovation intensified, driven by the desire to improve the effectiveness of the process by reducing waste. The new model also encouraged a more holistic perspective and recognized the system effect within the process. It also enhanced integration and knowledge exchange across the innovation process. The key contributions of this model to our perspective on innovation management were the increased engagement of external and internal stakeholders in the process, recognition of the complexity resulting from the relationships of parallel activities, and recognition that process effectiveness could be enhanced through ongoing analysis and learning.

FIFTH GENERATION

The fifth generation can be viewed as an extension of the fourth generation, where greater focus is placed on networking, system integration, and agile communication infrastructure. With the emergence of the Internet, globalization, and the concept of open innovation, the fourth-generation model has evolved toward a new generation. Today's innovation model must encompass innovation distributed across geographically dispersed organizations and yet maintain the integration that ensures that systems remain lean and efficient. The modern model must be capable of structuring the engagement of all relevant stakeholders and yet remain agile enough to adapt to contingencies. The fifth-generation perspective reflects the systemic nature of the innovation process, where interrelationships and feedback loops reflect the process complexity. The management of innovation reflects this interwoven nature, as opposed to the simplistic linear perspective of the earlier models. The implications for innovation management are that routines should be developed that nurture rather than constrain innovation development and that a balance between structure and flexibility should be maintained.

The various generational models have contributed to our understanding of the functioning of the innovation process and how it can best be managed for the benefit of the organization. Although all models have their value, the earlier models have left organizations with a partial perspective of the innovation process that is myopic and limiting. These limitations of the innovation process include the following (adapted from Tidd, Bessant, & Pavitt, 1997):

- A technology push that leads to an overemphasis of R&D discovery at the expense of other phases
- Concentration on radical and technological developments that ignores other types of innovation
- A focus on output rather than the process, which makes it difficult for organizations to manage and learn
- An inability to acknowledge the complexity and interdependence of the process because of a simplistic linear perspective
- Organizations defining certain occurrences within the process as random and then failing to adequately understand their root cause

It is evident that organizations that rely on a partial perspective of the innovation process and concentrate on managing only part of the process cannot achieve the same effective management. Organizations that understand all five generations of innovation have a better understanding of the holistic process and its complexity and thus are better able to manage and nurture innovations through the process.

Managing Innovation

Innovation management is the process of managing innovation within an organization. This includes managing ideas, goals, projects, and initiatives, improving communications, and managing innovation teams. Today's innovations will become obsolete in the future, and for organizations to sustain and adapt their mission they must learn how to innovate and replace existing products, processes, and services with more effective ones. Continuous innovation and learning are important to prevent organizations from repeating mistakes of the past and to improve their ability to exploit prospective opportunities. Our understanding of the innovation process and its management has evolved from a linear technology push model to a system integrating and networking model (Rothwell, 1994). The success of an innovation is influenced by factors such as the inherent characteristics of the innovation, the external environment into which it will be introduced, the organization's supporting infrastructure, the characteristics of potential adopters, and their perceptions of the innovation and the organization (Franklin, 2003). Innovation management is the process of managing information, people, and technology associated with innovation to influence the eventual outcome. It relates to the plans and routines the organization has developed to nurture an innovation from its birth as a creative idea through to an eventual delivery to the market.

PLANS

Innovation is often described as being either planned or unplanned. Unplanned innovation results from unplanned events such as problems that suddenly arise or a flash of inspiration. Planned innovation, on the other hand, results from a plan in which organizational goals are set and soon followed by actions such as idea generation aimed at meeting these goals. Both planned and unplanned innovations can be incorporated into one innovation plan. For example, goals contain many specific targets that must be achieved. Some goals can be defined that allow experimentation and, in effect, unplanned or serendipitous discovery. The activities in this book combine to create one innovation plan for an organization.

ROUTINES

Although each innovation is unique in its own right, if one examines the process by which innovation occurs in any organization rather than the innovative output, then certain patterns or routines begin to emerge. Identifying, understanding, and manipulating these routines provide the organization with a means of planning the development of its innovations from creative concept to market success. Much of this book focuses on creating simple routines for practicing some of the ideas in innovation management. Routines such as creating ideas, managing projects, and commercialization are a helpful way to elucidate the innovation process. The goal of this book is to present a set of routines that can act as a roadmap for applying innovation in any organization. The routines presented in this book are built around the core activities of defining goals, managing actions, empowering teams, monitoring results, and building communities.

Routines vary greatly between organizations. The routines developed throughout this book will not be adopted by all organizations. For example, some organizations do not adopt the practice of strategic planning, others do not practice project management, and others still have no performance measurement system. This book brings key routines together into one overarching process called the innovation funnel that contains routines for integrating goals, actions, teams, and results. Many well-established authors abhor attempts to create these types of routines, and for good reason. Routines can create safe structures that ultimately stifle free thinking and creativity, which often lies at the heart of most radical innovations. We encourage you to take this warning on board when reviewing the lessons in this book. However, routines do provide a framework that can be learned and applied by organizations. Simply viewing the innovation process as a black box may provide unlimited scope for developing innovative concepts, but it does not provide guidance or structure to help organizations manage and improve the process to enable them better to transform creative concepts into market successes. The routines presented in this book are designed to provide organizations with a foundation for developing structure for managing their innovation process. Over time, an organization can adapt these routines to better suit its unique contingencies and allow it to achieve more effective management of its innovation process.

EXAMPLE: The difference between a good mission statement and a bad one is very often in the eye of the beholder; that is, it depends on the person interpreting the mission statement. When a quality team comes together and agrees on a mission statement, to them it is a great mission statement. The greatness often comes from the process or from discussing and agreeing to the statement—the sense of sharing and common purpose. Often the words end up meaning significantly less than the shared experience, but they do remind participants of their shared experience, which is why they appear to be good statements for the participants and perhaps poor statements for those who did not participate. In the examples listed in Table 2.3, which mirror the organizations presented in Table 1.1, there are many good and bad mission statements. However, they all have one thing in common: They have been created and agreed to by the team behind each innovation plan. Study these cases, then go online and study the many excellent guides for creating an effective mission statement. Keep one thing in mind when looking into mission statements: the scope. Precisely whom is the mission statement referring to? For example, if your organization is a department within a larger multinational organization, then your mission can be significantly different from that of the larger organization, and perhaps your mission statement can refer to your own internal processes and customers.

Summary

Innovation is about making changes. Change management is a broad discipline that can incorporate many associated techniques such as project

Table 2.3   Mission Statements

Organization	Scope	Mission Statement
Reflective Display Films	Management team	To provide optical solutions to the display industry by using creative design and development techniques innovatively
Galway Tables	Management team	Meet our customers' changing needs by supplying high-quality products while continuously striving to improve lead times and awareness of the environment
Beglin Entertainment Centre	Management team	Provide a total entertainment experience that satisfies our customers in quality and value for money
Connemara Distillery	Management team	Delight customers with quality Irish whiskey
Qualtrans Translations	Management team	Language solutions for the global medical industry
Car Consultants	Management team	Giving car owners helpful and practical advice
Dynocorp Nightclub Group	Management team	The standard for nightclubs
HarPer Sculpting	Management team	Improving quality of life
Medical Device Manufacturing	Computer service department	Continuous improvement plan
Precision Engineering	Engineering department	Efficiency and quality through excellence in teamwork
Ratio Technology	Design department	Exceptional product value, large-scale customization, and reliable quality
Moneypits Bank, High Street Branch	Management team	Trusted member of the community
Children's Hospital	Engineering department	Serving our customers' engineering needs
Ambec Resorts	Single-hotel management team	A home away from home

management and knowledge management. These techniques have one thing in common: They promote and facilitate change in the organization. Many techniques such as strategic planning, performance measurement, project portfolio planning, and knowledge management have application for the management of innovation. Excellent organizations have developed traits that can be transferred to all organizations. Traits such as a bias for action and remaining close to the customer are universally important for all types of organizations. Culture is the bedrock on which all innovative efforts thrive or fail. There are a number of recognizable traits for changing toward a more innovative culture. Organizations need to create a learning culture that allows individuals to take risks, conduct experiments, implement actions, and, most important, reflect on the lessons learned. The organizations that will continuously innovate in the future will be the organizations that discover how to tap people's commitment and capacity to learn at all levels in the organization and develop routines that support the management of the innovation process. In the next chapter, we will examine the innovation process in more detail and see how organizations can create repeatable routines that can help to manage it.

Activities

Create a simple statement of 5 to 12 words that describes the overall mission of your organization. Search online for a definition of the term "organizational mission." This will help you understand the meaning of the term *mission*. A second part of this activity is to create a list of real competitor organizations that operate in the same sector as your organization. Go online and identify three to five suitable organizations that you will compete against for market share. As in reality, analysis of these competitor organizations can act as a stimulus for generating ideas for new products, processes, or services or simply provide interesting insights into the way they manage innovation. Copy Table 2.4 into a spreadsheet and complete the fields.

STRETCH: Open a research file on each of your competitors, analyze their performance, and track their innovation developments. Try to get your hands on their annual shareholder reports.

REFLECTIONS

- Explain the relationship between change and innovation.
- In his change management method, Kotter talks about creating a sense of urgency. Explain what this means.
- List a number of traits of excellent organizations.
- Explain some of the main traits of innovation culture.
- List a number of barriers to innovation.
- What is the fifth-generation innovation model?
- Explain the dangers of having too much routine.

Table 2.4   Competitors

Competitors		
Group	**Title**	**URL**

Group: Label of the group of links (e.g., "Competitors")

Title: Title of the company

URL: Link address

Processing Innovation 3

E very organization invests in innovation in order to change. Organizations put aside a proportion of turnover to change products, processes, and services. Particular objectives must be achieved because of this investment. However, a very large percentage of innovation activities fail to meet these objectives. The reasons behind failure give us clues about how avoid such failure in the future. In this chapter we examine the process by which organizations apply innovation. By understanding the process by which innovation takes place and then improving and mastering that process, organizations can lower innovation failure rates and speed up the process of growth. We also look at the process of innovation from idea generation, through evaluation, and on to realization, where customers become the ultimate judge of the success of an innovation. We conclude by looking at the innovation funnel that brings together four of the key ways to improve the ability of any organization to manage its innovation process: goals, actions, teams, and results.

LEARNING TARGETS

When you have completed this chapter you will be able to

- Show the main reasons why organizations invest in innovation
- Understand some of the reasons why innovation fails
- Discuss the key stages in the innovation process
- Understand the importance of opportunity recognition in the innovation process
- Explain the innovation funnel
- Apply an innovation method to building an innovation plan

Innovation Investment

Each year organizations spend a significant amount of turnover on innovation. The amount of investment can vary from as little as 0.5% of turnover for organizations that operate in stable marketplaces to more than 20% of turnover for organizations in emerging or turbulent marketplaces. The level of expenditure depends on the aspirations and ambitions of the individual organization and whether it has growth potential. The average expenditure across organizations is just under 4% of annual turnover (European Commission, 1996). For an organization with a turnover of $1 billion, this represents an annual investment of approximately $40 million. This budget typically is spread across various functions, including product design, information systems, manufacturing systems, and quality assurance, to allow innovative actions to be undertaken. As the innovation budget is often based on a percentage of forecasted turnover, three potential outcomes are possible. First, forecasts are correct, and thus the allocated budget is also correct, allowing planned innovation to be undertaken; second, the actual turnover exceeds the forecast, resulting in a budget that allows a greater number of innovative initiatives to be undertaken; and third, the actual turnover is less than that forecasted, resulting in insufficient budget to undertake the planned innovative projects. The latter scenario can result in an organization merging together, postponing, or even abandoning innovative projects because of the increased financial constraints. One disadvantage of this mode of innovation investment is that when an organization's turnover is decreasing, this is perhaps the time when the investment percentage should be increased. Innovation investment can be linked to more strategic rather than operational results. The topic of investment in innovation is discussed in further detail in Chapter 10.

Goals of Innovation

The principal goals required by an organization in return for this investment vary between organizations. The following have been found across a large number of manufacturing and service organizations and ranked in order of popularity, with the first goal being common to most organizations (European Commission, 1996):

1. Improved quality

2. Creation of new markets

3. Extension of the product range

4. Reduced labor costs

5. Improved production processes

6. Reduced materials

7. Reduced environmental damage

8. Replacement of products or services

9. Reduced energy consumption

10. Conformance to regulations

The first goal suggests that the most common reason for organizations to invest in changes to products, processes, and services is to improve quality. Most of these goals range across improvements to products, processes, and services and dispel a popular myth that innovation deals mainly with new product development. Most of the goals could apply to any organization, be it a manufacturing facility, marketing firm, hospital, or local government.

Failure of Innovation

Reaching particular goals is the ultimate objective of the innovation process. Unfortunately, most innovation fails to meet organizational goals. Failure rates vary widely depending on the type of innovation being undertaken, the experience level of those undertaking the action, and the particular context in which it is being implemented. Research cites failure rates of 50% relating to achievement of planned goals (Strebel, 1999), with other research claiming failure rates of up to 70% for new technology projects and 80% for new process initiatives (Burnes, 1996). From another perspective, a survey about product innovation highlights the fact that out of 3,000 ideas, only one will become a success in the marketplace (Stevens & Burley, 2003). As a consequence of the turbulent nature of innovation, a certain level of failure is an inevitable part of the innovation process and directly related to the level of risk the company is comfortable exposing itself to. All organizations experience failure, but the successful ones choose to monitor or understand why it has occurred and identify what can be done in the future to prevent it. The impact of failure goes beyond the simple loss of investment. The cost of failure can also include loss of morale among employees, increased cynicism, higher resistance to innovation in the future, and loss of lead time over competitors.

Innovations that fail are often potentially good ideas but have not been properly exploited by organizations because of budgetary constraints, lack of skills, poor management, or poor fit with the organization's current goals or market requirements. Failures will never be totally eradicated from the innovation process; rather, they should be identified and screened out as early as possible. Organizations that try to avoid failure completely will inhibit the level of creativity of employees and skew their portfolio of innovations toward incremental changes. Early screening prevents unsuitable ideas from devouring scarce resources that are needed to advance more beneficial ones. Organizations can learn more about failure when it is openly discussed and debated. Lessons learned from failure often reside longer in the organizational consciousness than lessons learned from success. Although learning is important, high failure rates throughout the innovation process are wasteful and should be monitored to avoid mistakes being repeated.

The causes of failure can vary widely depending on the individual innovation. Some causes will be external to the organization and outside its influence of control. Others will be internal and ultimately within the control of the organization. Some of the more common causes of failure in organizations can be distilled into the following five types (O'Sullivan, 2002):

- Poor goal definition
- Poor alignment of actions to goals
- Poor participation in teams
- Poor monitoring of results
- Poor communication and sense of community

Poor goal definition means that organizations find it difficult to define their goals. Poor goal definition requires that organizations decide on appropriate goals for their environment and define these goals in terms that are understandable to everyone involved in the innovation process. Poor alignment of actions to goals means that organizations find it difficult to continuously link the ideas and projects they are pursuing with their goals. This is perhaps even more acute if goals are difficult to define in the first place. It also influences effective management of the portfolios of projects that the organization is undertaking to ensure they are balanced appropriately. Poor participation in teams refers to the behavior of individuals and teams, latent knowledge of the organization, and the particular skills of individuals to contribute to the achievement of innovation. It also refers to the payment and reward systems that link individuals to goals. The poor monitoring of results refers to sharing of the status of goals, actions, and teams involved in the innovation process within the innovation team and its main stakeholders. Finally, poor communication and sense of community relate to ineffective channels of communication and collaboration that constrain knowledge sharing and the ability of employees to participate as a broader community in the innovation process and make informed decisions when needed.

Process of Innovation

The ability to manage the innovation process is an essential competence of any organization, but members must first understand the workings of the process to be successful. The path innovative concepts follow from their initial generation as ideas through to their eventual consumption by the intended market can vary greatly (Tidd et al., 2005). In this and subsequent sections we will present two complementary ways to understand the innovation process. In this section we present the process as comprising four interacting processes, from idea generation through to eventual realization. This is a broad definition of the innovation process. Later we will present the innovation process as comprising five key knowledge areas that can be easily translated into a knowledge management system for innovation. This latter definition is based on the innovation funnel and focuses strictly on the aspects of innovation that can be codified into a set of simple innovation tools.

The process of innovation can be described in terms of the interactions between four key subprocesses (Figure 3.1):

- Idea generation
- Opportunity recognition
- Development
- Realization

Two related subprocesses are associated with opportunity recognition: organizational goals and available resources. In addition, there is another subprocess that is not illustrated in Figure 3.1 but underpins all processes: learning. The learning process permeates each of the processes, from idea generation to realization.

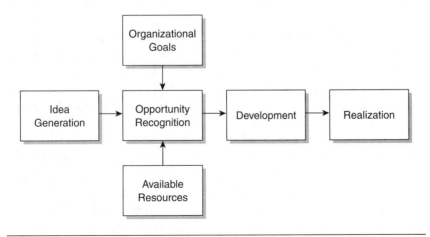

Figure 3.1   Innovation Process

IDEA GENERATION

The first stage in our perspective of the innovation process relates to the creative activity of generating an opportunistic idea. This stage involves the continuous scanning of the internal and external environment for threats and opportunities that might be developed into an innovation by the organization. This stage involves mining the sources of innovation for new ideas and evaluating solutions to identified problems. An organizational culture that encourages creativity and empowerment can significantly support this phase of the process. The input typically stems from a technical insight into a product or process or thoughts about a service. In some cases ideas arise from observed problems that have occurred in the past or may occur in the future. Ideas can also be stimulated by the goals of the organization or an unanticipated opportunity. Various stimuli can lead to the creation of an idea and range from reading magazines and observing problems to visiting other organizations and having informal discussions with colleagues and customers.

OPPORTUNITY RECOGNITION

The second stage of the process is opportunity recognition, in which the opportunity of developing the idea into a new product, process, or service is assessed and evaluated relative to other opportunities. This phase of the process involves deciding which innovative ideas will be pursued by the organization and which are deemed outside its interest. The undertaking of innovative actions is both expensive and resource intensive for any organization, and even large organizations such as 3M and Intel need to choose which ideas to pursue. How this decision is made can be complex and involves tradeoffs, including correlation with the strategic goals and resources available to the organization, the organization's current capability, the mix of innovations already being developed, the actions of competitors, and the emerging signals from the external environment. Similarly, this evaluation of prospective innovations is not a onetime event but occurs periodically during the innovation process to ensure that the organization is investing in positive innovations. Cooper (1986) refers to these decision points as "stage-gates," where unsuitable initiatives are eliminated to allow extra resources to be directed toward more suitable innovations. Two types of error can occur at this phase of the process: An idea that would have been successful for the organization may not be pursued, or an idea that will be unsuccessful for the organization may be allowed to continue. The more damaging of these errors is the latter because the development of this idea will consume scarce resources and prevent another beneficial idea from being developed. In scenarios where a good idea is wrongly abandoned, it is likely that in a supportive culture, this idea will recur at the idea generation phase. The difficulty in this phase of the process is that the organization does not have a crystal ball to see into the future and therefore cannot know for certain which ideas will be winners or losers. Members of the organization can only make the most enlightened decision they can, based on available knowledge, and continue to periodically screen their portfolio of developing innovations for appropriateness. As a consequence of this phase, ideas are often improved, merged with other ideas, or in many cases shelved or abandoned. An important test for an idea is that it match the goals of the organization and available resources, such as people and money.

DEVELOPMENT

If an opportunity is recognized as appropriate for the organization, then the idea moves to a new stage where it can be developed further. This phase involves the development of the idea or solution into a potential innovation that is ready for launch to its internal or external market. The development of an innovation can be highly resource intensive for any organization. The selection of innovations by an organization is constrained by the budget and the existing portfolio of innovative actions. Similarly, certain innovations may require competencies and skills that are scarce or even absent from the organization, and this scarcity can hinder the implementation of certain innovations. Organizations must carefully manage the innovative actions, ensuring that they are adequately resourced to ensure success. Part of managing the implementation of these actions is constant scanning of the external environment for emerging trends that may alter the trajectory of the innovation. The development phase of the innovation is usually undertaken as a team approach (because of the diverse competencies needed) and involves making the initial idea tangible in a form that best meets market demands. Key activities of this phase can include experimentation, design and development, testing, market analysis, and prototyping. At the end of the development phase, the initial idea has been developed into a tangible product, process, or service that the organization views as capable of meeting user needs. Many potential innovations wait at the end of the development phase for market conditions to be right before they move to the realization phase.

REALIZATION

This phase of the innovation process relates to the launch to the market, which is where the customer makes the final evaluation of the innovation. Understanding customer needs is essential to ensure that the eventual offering to the market meets these needs. A strong alignment between the objectives of the particular innovation and the needs of the customer increases the likelihood that the innovation's initial market adoption will be a success. This fact becomes most pronounced with respect to technology innovations, where the organization must manage fulfillment of each of the customer segments across the product life cycle (Moore, 1999). Although Figure 3.1 represents the realization phase as following the development phase, in reality these phases overlap. Market information about customer needs is an essential input to the development phase, and information about the innovation's attributes is necessary to begin educating and preparing the marketplace. The objective of the realization phase of the process is to develop an innovation for the market that meets customers' needs and is readily adopted. When the organization is developing a process innovation, the market can be said to be internal. Consequently, the realization phase encompasses activities such as commissioning, validation, and training to facilitate its successful adoption.

LEARNING

Learning is the final subprocess in the innovation process. It requires the organization to analyze the previous phases of the innovation process and identify areas where the process can be improved. In this way, even innovative actions that are abandoned or end in failure can be beneficial because the organization can learn from its mistakes and avoid repeating them in the future. Similarly, the new knowledge acquired from undertaking the prospective innovations can also be used as input to the idea generation phase that may lead to future innovations. Over time the organization's effectiveness at managing its innovation process improves, which will also increase the success of its future innovative actions.

Applying Innovation

Every organization would like to be able to increase the success of its innovative efforts in order to enhance its competitive position for the future. Although many leading organizations have invested significant resources in developing the culture and routines for their innovation processes, most organizations continue to rely on the efforts of a handful of people and chance. An innovative organization is one that can perfect these routines in addition to creating an innovation culture in the organization that engages people. Five key routines can facilitate its management of the innovation process (Dooley & O'Sullivan, 2003). These mimic the five root causes of failure discussed earlier:

- Better definition of goals
- Better alignment of actions to goals

- Greater participation of individuals in teams
- Better monitoring of results
- Greater communications and building of communities

GOALS

The term *goals* refers to the objectives that the organization wants to achieve by engaging in innovation. The organization needs to decide on the goals it will pursue in the future that will enhance its competitive advantage. Defining these goals provides the innovation trajectory for the organization and is a key factor in creating high-impact innovation. We will deal with goals in more detail in Part II. There are a number of ways of defining goals, including the following:

- Statements such as the mission and vision statement
- Needs of stakeholders such as customers and shareholders
- Objectives such as strategic plans
- Indicators of performance such as output and profits

ACTIONS

The term *actions* refers to the expenditure of effort on the part of the organization in developing creative concepts into eventual innovations. Organizations that possess routines that nurture the flow of actions across the various phases can enhance the effectiveness of their innovation process. A key issue is that actions are in some way aligned with the goals of the organization so that they contribute to developing the organization in the direction of its defined goals. We will deal with actions in more detail in Part III. Actions include the following activities:

- Identifying problems and solutions
- Generating ideas
- Managing initiatives and projects
- Managing project portfolios

TEAMS

The term *teams* refers to the resources used for the innovation action. Teams are made up of people who use their skills and other organizational resources, such as finance, to facilitate the development of innovative actions. The more people engaged in the process, the greater the creative capability and skills available to support the innovative actions. A key part of every team is the available resources in terms of time, knowledge, and equipment, but particularly funding. The issues of funding are dealt with in a number of areas in the book, particularly in Part III. We will deal with teams and aspects of leadership, team performance, and appraisal in more detail in Part IV. There are a number of issues related to achieving greater participation by individuals in teams, including the following:

- Team leadership
- Building structure in teams
- Improving participation by individuals
- Linking the performance of individuals to organizational goals

RESULTS

The term *results* refers to the outcomes of an effort to innovate. There are clearly many things going on at once, and organizations must learn to use techniques that allow them focus on critical activities. The importance of monitoring results allows the organization to assess its ongoing progress and, if necessary, alter the mix of actions it has flowing through the innovation process. We will deal with results in more detail in Part V. The principal results that an organization needs to concern itself with are the following:

- Results of goals such as objectives and indicators
- Results of actions such as ideas and projects
- Results of teams such as where individuals are participating and how their performance review is progressing

COMMUNITIES

The term *communities* refers to all people who share a common purpose of supporting innovation in the organization. This community is often motivated by the defined goals of the organization but can also be influenced by the professional goals of specific individuals. As innovation has become more complex, community has broadened to encompass people in other organizations who are engaged in the innovation. We will deal with communities and community-related issues throughout the book, particularly in Part V. Building community is a time-consuming process and involves key issues such as organization and leadership, communications, and knowledge management.

Innovation Funnel

The innovation funnel provides a solution for effectively managing innovation by controlling the interaction of goals, actions, teams, and results used in the innovation process (Dooley & O'Sullivan, 1999, 2000). The funnel illustrates how goals, actions, teams, and results interact with each other to deliver innovation in any organization (Figure 3.2). The funnel metaphor is not new and can be traced back to the seminal work of Hayes, Wheelwright, and Clark (1988) in relation to their work on the development funnel. The innovation funnel can be visualized as containing four arrows flowing around a funnel. The arrows represent the flow of goals, actions, teams, and results. Actions enter the wide mouth of the funnel and represent, among other things, ideas and potential solutions to problems. These actions flow toward the neck of the funnel, where they are screened in terms of their opportunity for the organization. The neck of the funnel is constrained by two arrows: goals and teams. These constraints loosen or tighten depending on the availability of teams and the goals defined by the organization. Tightly defined goals can be visualized as closing the neck of the funnel, resulting in fewer actions flowing through. The availability of more teams (increased skills or funding) can be visualized as opening the neck of the funnel and allowing more actions to be developed. As actions are implemented, they affect the organization's fortunes. The final arrow reflects this impact. Results flow from the narrow end of the funnel and represent information about the status of goals, actions, and teams and their relationships. This arrow flows back toward goals, representing the impact of results on the process of defining and redefining goals.

An important aspect of the innovation funnel is the relationships between goals, actions, teams, and results. For example, ideas that cannot easily be related to goals will find it difficult to proceed into the funnel. This has two effects. First, the individuals or teams generating the ideas will study the goals more closely in order to generate an idea that matches better. Second, good ideas that are not easily associated with goals will begin to affect the definition of the goals, perhaps ultimately leading to a redefinition

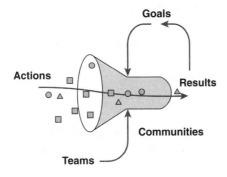

Figure 3.2   Innovation Funnel

of goals in order to allow the good ideas through. This is a natural learning process in an innovation community. When goals change, the generation of ideas that meet these goals increases because the innovation community is now tuned to having new ideas meet the new organizational goals. The process offers the innovation community the ability to change the innovation process in response to changing demands of stakeholders.

Innovation Knowledge

The innovation funnel illustrated in Figure 3.2 can be expanded to include examples of the many ways in which goals, actions, teams, and results can be defined and codified in a simple knowledge management system. The approach adopted in this book is to use simple tables to store and share innovation knowledge. This can very easily be expanded to a sophisticated online knowledge management system. Figure 3.3 illustrates an expanded version of the innovation funnel highlighting a number of unique knowledge elements. Think of each word in this diagram as a worksheet or table in a spreadsheet or as a software module or web part in a knowledge management system.

For example, goals can be defined through knowledge elements such as statements (e.g., a mission statement) and indicators. Actions are defined through elements such as problems, ideas, and projects. Teams are defined and codified through individuals and teams but also through performance appraisals. Results are defined through exceptions and reports. Finally, communities can be defined and codified through knowledge elements such as notices, blogs, and libraries. All of these knowledge elements combine and interact with each other to create a sophisticated knowledge management system for managing innovation in any organization.

Most organizations do not implement all of these knowledge elements at once. They choose only the elements that are relevant for their own particular organization at a particular moment in time. Over time, as the organization builds the various knowledge elements, the power of the system increases as elements interact with each other to illuminate a more holistic innovation management process. This particular innovation funnel will be implemented as part of the activities at the end of each chapter in this book. The remainder of this book will look at many of these knowledge elements in greater detail.

EXAMPLE: Clearview Pharmaceuticals is a small manufacturing company. The innovation team is responsible for mainly process innovations and includes key personnel from all of the main functions in the organization: all managers and some specialists. They meet bimonthly to discuss the progress of their goals and review the status of various projects. They also review any ideas that have been generated by employees that match organization goals. Clearview Pharmaceuticals' mission is "the world-class manufacturing of eye care products and engagement of employees in

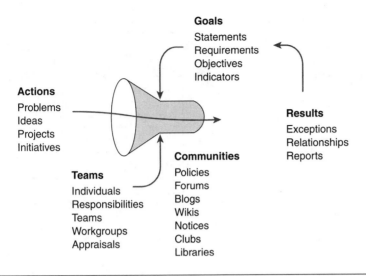

Figure 3.3   Innovation Funnel Expanded

continuous process innovation." Product design is the responsibility of a sister organization at another site. The team has identified four major drivers of innovation: conformance bodies, which impose the need for greater traceability in the production process; their parent corporation, which regularly demands cost reductions; customers, who want greater flexibility and shorter lead times; and suppliers, who expect greater forecasting accuracy in return for shorter lead times and lower costs. The innovation team, listed in Table 3.1, has developed an innovation plan titled "Manufacturing Development Plan 2007–2010." The plan outlines the goals for the 3-year period, a set of performance indicators that are monitored weekly, a dynamic set of projects, and the responsibilities of each person in terms of goal attainment and project management.

Summary

Organizations spend on average just under 4% of turnover on innovation, trying to achieve goals such as better quality, lower lead times, more product variety, and increased market share. Most innovation fails to achieve these goals, and some analysts argue that failure could be as high as 80%. The causes of failure are varied, but some common causes of failure can be found that affect most organizations. These causes can be divided into cultural and process failures. Cultural failures such as poor leadership and organization are clearly important but take time to improve. Process failures such as poor definition of goals and poor alignment of innovative actions with goals are also important but can be remedied in the shorter term through better team behavior and better management of innovation knowledge. Irrespective of

Table 3.1   Innovation Team at Clearview Pharmaceuticals

Individuals	
Name	**Job Title**
Andy Scott	Management Consultant
Colm Griffen	Logistics Manager
Dan Hyland	Marketing Manager
Dave Mahon	IT Manager
Gary Smyth	Senior Engineer
Gutz McFadden	HR Manager
Kevin Staunton	Health and Safety Officer
Mark Ryan	Finance Controller
Mary Joyce	Plant Manager
Mary Kelly	Material Manager
. . .	

the success of an innovation, over time it will be superseded by other innovations. The organization must be able to manage its innovation process effectively if it is to deliver ongoing innovations in the future. The innovation funnel offers organizations a structured approach to managing innovation that reduces the effects of some of the key causes of failure while simultaneously facilitating goal attainment. The rewards for adopting such a simple yet effective system can be significant not only in terms of costs and benefits but, more important, in terms of morale and skill development among participants in the innovation process. As the organization operates the innovation funnel, it identifies areas for improvement and thus can enhance its ability to innovate. The ability to learn faster than your competitors may ultimately be the only sustainable competitive advantage. Creating an innovation that gives an organization temporary competitive advantage may result accidentally or from the actions of people who ultimately move on to other organizations. Sustainable competitive advantage requires that organizations master the management of their innovation process so they can continuously innovate in the future.

The remaining four parts of this book explore the innovation funnel in detail. Their respective chapters are illustrated in Figure 3.4. Part II, "Defining Innovation Goals," contains three chapters that discuss a number of aspects of applying innovation, including how to analyze the environment of the organization, define strategic objectives, and deploy performance indicators. Part III, "Managing Innovation Actions," contains four chapters that look at a number of applied tools and techniques used for creativity, project management, new product development, and project portfolio management. Part IV, "Empowering Innovation Teams," contains three chapters that explore the application of innovation in areas such as leadership, teams, and performance appraisal. Part V, "Sharing Innovation Results," contains three chapters that look at the application of technologies and techniques for managing knowledge, building communities, and extending innovation beyond the boundaries of a single organization.

Activities

This activity requires you to populate your organization with a team of people who will be responsible for developing your innovation plan. If you have chosen a large organization, then the team will be primarily senior managers and some specialists. If your organization is a small department, then it may be every member of the department plus a few key people from other departments. For your organization, define approximately 10 to 16 employees in your organization. These people will participate in developing and implementing the innovation plan you will develop for your organization. As you progress through the activities, you can revisit this activity and add new employees and functions as needed. Name each person and define his or her role in the organization (e.g., senior technician, general manager). You are also encouraged to define the various functions in your organization and any relationships between them. Copy Table 3.2 into a spreadsheet and complete the fields defined.

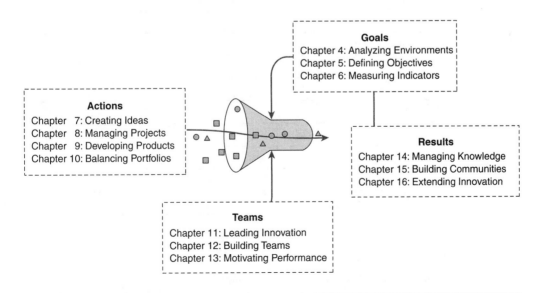

Figure 3.4   Book Chapters

STRETCH: Another element of this activity could be to create an organizational chart. This can be a functional or department chart that identifies people and their responsibilities and reporting structure to each other.

REFLECTIONS

- List some of the main reasons why organizations invest in innovation.
- What are the key stages in the innovation process?
- Explain the opportunity recognition stage of the innovation process.
- What are main causes of failure to achieve innovation?
- What is applying innovation?
- Explain the innovation funnel.

Table 3.2   List Your Team

Team	
Name	**Job Title**

Name: Names of team members (e.g., John Doe)

Job Title: Job title or skill title (e.g., production supervisor)

Part II

Defining Innovation Goals

D efining innovation goals is the first and perhaps the most important activity in creating an innovative organization (Figure II). Well-defined goals inspire ideas and guide activities toward the achievement of performance targets. An organization defines its goals in a number of different ways. These include statements, stakeholder requirements, strategic objectives, and performance indicators. This part of the book looks at ways of defining goals. The first chapter looks at the goal definition process. The roadmap outlines a number of steps that need to be taken to develop a comprehensive set of goals. Chapter 5 looks at defining stakeholder requirements that are used to inform the strategic planning process. The demands of customers and shareholders are an important input to goal definition. Chapter 5 also looks at the strategic planning process in detail. Key steps in strategic planning include identifying strategic thrusts and defining appropriate strategic objectives for the organization. Chapter 6 examines the importance of performance indicators and the role they play in communicating goals to every person in the organization and in fostering innovation. Collectively, these types of goals integrate to form a comprehensive portfolio of innovation goals for any organization.

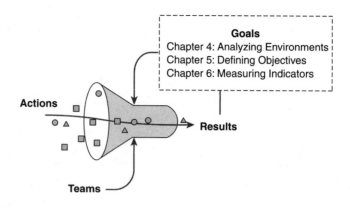

Figure II   Defining Innovation Goals

LEARNING TARGETS

When you have completed this part you will be able to

- Outline a process for defining innovation goals
- Use techniques for conducting environmental analysis
- Understand the importance of benchmarking in innovation
- Explain how to develop statements such as mission and vision statements
- Identify stakeholders and their requirements
- Describe the process of strategic planning
- Outline the process of creating performance indicators

Analyzing Environments 4

U nderstanding how an organization fits in its external environment is essential to applying innovation. The ability of the organization to adapt to meet the emerging needs of the environment is the basis of its survival and the source of many of its innovative ideas. Every organization has its own way or method of defining goals that will better align it with the requirements of the emerging environment. In this part we explore four main types of goals. These goals are called statements, requirements, objectives, and indicators, and all influence the types of innovative actions pursued by the organization. An important first step in deciding on these goals is to analyze the external environment. In this chapter, we look at environment analysis that informs the goal definition process using techniques such as political, economic, sociocultural, and technological (PEST) analysis; benchmarking; and strengths, weaknesses, opportunities, and threats (SWOT) analysis.

LEARNING TARGETS

When you have completed this chapter you will be able to

- Outline the process of defining goals for an organization
- Understand the five-forces model used in business analysis
- Explain the difference between rational and incremental goal definition
- Explain the benchmarking process
- Understand the PEST process for environmental analysis
- Develop SWOT statements for innovation
- Begin the process of codifying critical data for innovation

Goal Planning

Goal planning is a common technique in most organizations. This was not always the case. In the past, goals resided in the minds of owners and senior executives, who communicated them to subordinates through verbal instructions. Today's complex organizations need employees to understand and share common objectives in order to fully engage in idea generation and project execution. Few owners and managers now have the ability to set goals for an organization and simultaneously generate and manage the associated actions needed to make change happen. The principal approach of goal planning is to define high-level goals for innovation within the organization and for these goals to guide individuals and teams in generating actions such as ideas and projects. The goals of the organization inform the imagination of individuals and set the performance indicators that the organization needs to achieve. The organization achieves this movement from its current to its desired future position through the innovative projects and actions it undertakes.

Goals are general and high level because they typically guide development for 1 to 5 years into the future. They are also general because they need to leave room for individuals to make their own decisions when making specific proposals on how these goals can be achieved. The process by which an organization transforms over time is sometimes called strategic management. The strategic management process strives to identify the best position for the organization to occupy in the future and determines the most suitable routes by which it can move there. The strategic management process consists of three interrelated elements: strategic analysis, strategic choice, and strategic implementation (Johnson & Scholes, 1997). These elements combine to inform the strategic goals, the strategic actions for achieving these goals, and the implementation of these actions.

The approach to strategic management relating to innovation is illustrated in Figure 4.1. In this model, the key activities are statements, stakeholder requirements, strategic objectives, performance indicators, actions, and results.

All these activities are interrelated, and as the organization learns more about what impact each activity has, this can lead to changes to other activities. This framework presents a process whereby the current and future shape of an organization can be translated into a set of statements. These statements together with stakeholder requirements can then be translated into strategic objectives. These objectives in turn yield a set of performance indicators that ultimately motivate action and lead to results. The presence of this framework allows the link between executive-level plans and operational-level initiatives that will move the organization toward its vision. This part of the book deals with defining the goals of innovation and how they act as a guide for innovative projects and actions. Implementation issues regarding actions, teams, and results are dealt with in Parts III, IV, and V, respectively.

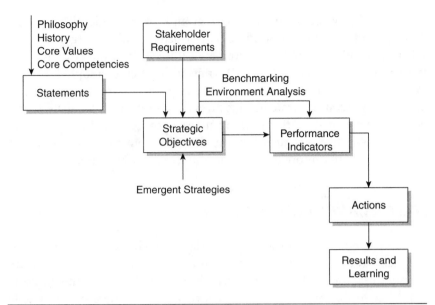

Figure 4.1   Goal Planning Process

STATEMENTS

Statements are high-level goals that inform individuals in the organization and their stakeholders about what the organization is doing or aims to do in the future. Typical statements include mission and vision statements. Mission statements are informed by the philosophy, history, core values, and core competencies of the organization. The vision statement is informed by the current and future environment in which the organization will compete. Statements in turn inform the process of generating strategic objectives. Statements can also include strengths, weaknesses, threats, opportunities, core competencies, quality, safety, and so on. There is more on statements later in this chapter.

STAKEHOLDER REQUIREMENTS

Requirements are expectations and demands from stakeholders about what they expect the organization to achieve or deliver. There are various stakeholders in an organization. Stakeholders can be seen as the customers in the broadest sense of the term. A key stage in setting goals is identifying who the stakeholders are and what they require from the organization. Their requirements inform the strategic objectives of the organization. There is more on requirements in Chapter 5.

STRATEGIC OBJECTIVES

Objectives are broad decisions about what the organization chooses to accomplish over a planning period. Strategic objectives are a more detailed list of goals, typically divided into groups of strategic thrusts. The development of strategic objectives is informed by both the organizational statements and stakeholder requirements. It is also informed by emergent technologies and state-of-the-art trajectories being adopted by the industry in general (e.g., "develop more e-business opportunities"). The process of generating strategic objectives is also informed by analysis and benchmarking of the organization's internal and external environment. Strategic objectives in turn inform the development of performance indicators. There is more on strategic objectives in Chapter 5.

PERFORMANCE INDICATORS

Performance indicators are the criteria by which the performance of a product, process, or service can be measured and evaluated. They are measurable targets of performance. They are linked with the strategic objectives in that each strategic objective should be measurable directly or indirectly through indicators. Selection of performance indicators is informed by such activities as benchmarking and environment analysis, as well as the higher-level goals (statements, requirements, and objectives) that have been chosen by the organization. The indicators selected by the organization will in turn inform the actions that the organization needs to generate, manage, and execute in order to move the organization toward its desired future position. There is more on indicators in Chapter 6.

EXAMPLE: The process outlined in this part identifies four main types of goals: statements, requirements, objectives, and indicators. Not all four are of equal priority for every organization. An alternative approach to defining goals is to adopt guidelines and standards such as ISO 9000 and the Malcolm Baldrige Award criteria. For example, manufacturing organizations use off-the-shelf systems such as world-class manufacturing, lean manufacturing, and total quality management to define their objectives and drive the determination of their performance indicators. It's generally accepted that all organizations, including banks, hospitals, and local governments, need to adopt a structured goal definition process.

Defining Goals

Many theorists have debated the value of defining goals. This has led to two different perspectives regarding goal development: the rational and the incremental (Tidd et al., 2005). The rational perspective views goal definition as a conscious effort to structure the development of an organization toward a planned future. The incremental perspective perceives goal definition as an ongoing stream of decisions regarding the development of the organization that eventually determines an organization's future (Burnes, 1991). Burnes proposes that both perspectives have merit and are contingent on the level of turbulence in the external environment. The higher the turbulence in an environment, the more difficult it is for an organization to define appropriate goals to move it into the future and the more likely it is that these goals will have to be changed to reflect emerging trends. Organizations need to develop goals that chart their development into the future but should be aware that these goals are based on certain assumptions and predictions that must be reevaluated for suitability as the future comes to pass.

Planning Period

Goals typically are developed at the beginning of a period of 1 to 5 years depending on the stability of the environment. Manufacturing and health-care organizations typically operate in an unstable environment and usually adopt plans of 1 to 3 years. Universities and government organizations typically operate in a more stable environment and can adopt plans of 3 to 5 years and sometimes even up to 10. Plans are reviewed periodically to ensure that they remain appropriate and reflect emergent trends in the organization's environment. Actions such as ideas and projects are generated, managed, and completed on a continuous basis throughout the planning period. At the beginning of the new planning period, actions from the previous planning period populate the new plan. However, as resources are released and new ideas are generated, new actions find their way into the new plan. Because of the continuous nature of the action process, some actions will remain uncompleted at the end of the planning period, and these are often continued into a newer planning period. The progress of goals and actions is monitored continuously throughout this period. Result monitoring not only informs the organization about how it is progressing but also helps with the process of assessing the appropriateness of the goals and actions. This learning activity perfects the goal planning process and aligns the organization with its external environment.

Environment Analysis

In order to develop appropriate goals to move the organization forward, one must first determine where the organization is. Environment analysis facilitates this endeavor and involves taking time to analyze the internal and external environment. In large organizations, consultants and special planning teams are often used to facilitate this process because of its complexity. The activity provides the innovation team with a valuable opportunity for learning and idea generation. By undertaking this type of analysis, organizations can develop goals that will steer them along the most appropriate path for the future. Many different types of tools and approaches are used to facilitate environmental analysis at both a macro and a micro (or industrial sector) level. These include PEST analysis, the five-forces model, competitor benchmarking, SWOT analysis, and understanding core competencies (Figure 4.2).

PEST Analysis

PEST is an acronym for political, economic, sociocultural, and technological and refers to the analysis of these influences on the organization (Johnson & Scholes, 1997). The PEST approach examines the environment in its widest context, assessing the macroenvironmental influences that will affect the organization going forward. Organizations need to assess the political situation that is likely to affect the organization going forward, including the impact of changes to factors such as employment and taxation law, openness of the economy, and the general stability of the government. Economic factors examined can include potential income available in the economy, prevailing interest rates, levels of unemployment, and stage of the business cycle. Sociocultural factors examined might include changes in demographics, lifestyle, and attitudes and beliefs of customers. Technological factors that might be examined include

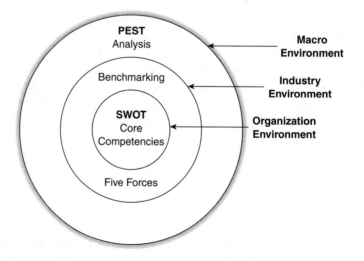

Figure 4.2   Levels of Environmental Analysis

changes in spending or the focus of R&D, relative public–private balance of R&D investment, and shifts in technology trajectories. Because these influences are at a macroenvironmental level, the majority of organizations rarely have a chance to influence these factors directly and can only react to them. The knowledge gathered by the organization regarding potential developments of these influences can provide valuable insights into the planning process and help identify the best position for the organization to be in.

Example: The following factors have been identified as part of the development of a new marketing campaign for a component supplier. The organization aims to expand into new Asian markets with existing products and services.

Political Factors

- Stability of the political environment
- Government policy and laws relating to trade and taxation
- Culture and religion
- Trade agreements

Economic Factors

- Interest rates
- Inflation and employment
- Gross domestic product

Sociocultural Factors

- Dominant religion
- Attitudes toward foreign products and services
- Impact of language
- Suffrage
- Population demographics

Technological Factors

- Cost of production
- Internet access and usage
- Mobile phones

Five-Forces Model

The most widely used environmental analysis technique is the five-forces model, developed by Michael Porter (1980). The framework analyzes competitive pressures present in the organization's industry sector under the following headings.

COMPETITIVE RIVALRY

Competitive rivalry between organizations creates pressure for change and is often the most recognized pressure by organizations. The level of this rivalry can vary across industry sectors and will influence the goals an organization chooses to pursue. Rivalry can range from the equivalent of "cozy cartels" where competitors avoid head-on competition to industry sectors where rivalry is fierce and organizations engage in price wars to lure customers away from competitors. In certain situations, rivalry can be so strong that organizations engage in below-cost selling in order to gain market share; clearly this scenario is not sustainable in the long term but can drive out weaker organizations from the market, resulting in increased share for those remaining. This behavior will have implications for the goals the organization pursues and the innovative actions undertaken to achieve them.

EXAMPLE: The level of rivalry within industry sectors can change over time. A good example of this is the European airline industry. In the mid-1980s, the sector was dominated by national carriers that operated in a highly regulated industry where there was limited access to routes and landing slots. With the deregulation of the sector in the early 1990s, competition increased dramatically, with low-cost airlines such as EasyJet and Ryanair entering the market. The impact of increased rivalry has been beneficial for the consumer, with more choice and lower costs, but has transformed the industry landscape, with a number of national carriers failing to adapt to their new environment and going out of business. This highlights the need for organizations to analyze the level of rivalry in their industry, both currently and in the future, in order to adopt the most appropriate goals to compete against rivals.

THREAT OF ENTRY OF POTENTIAL NEW COMPETITORS

The ease with which a new firm can enter the market influences the competition within the sector. New competitors not only increase competition by taking market share from existing firms but also often introduce new technologies, processes, and business models to the sector that can cause a disruptive shift in the status quo (Utterback, 1996). The presence of barriers to entry in a sector can act as a deterrent to new firms interested in entering the market because overcoming these barriers makes the proposition less attractive. Another factor that influences the threat of new entrants is the expected retaliation of existing organizations in the sector, which may use their size and economies of scale to force the new entrant out of the market again. Often the threat of such practice is enough to dissuade new organizations from entering the marketplace. However, if new entrants perceive that the opportunity is worth the risks, then they will incur the cost of overcoming the barriers to enter the market. Therefore, existing firms must be wary of taking excessive profits that make the market attractive to new entrants. One source identifies factors such as economies of scale, learning and experience curve effects, resource disadvantages, lack of a suitable technological platform, brand loyalty, regulation, and tariffs as some of the barriers that can prevent new firms from entering a market (Utterback, 1996). As was evident in the European airline sector, once the sector was deregulated, new entrants rushed to enter the sector and transformed the competitive landscape.

COMPETITIVE PRESSURES OF SUBSTITUTE PRODUCTS

The level of choice customers have in fulfilling their needs by adopting different types of products can have a significant impact on the competitive behavior of the industry. If the industry's product offering is one that has direct or close substitutes, then the actions of the alternative sector can affect an organization's market share and behavior. The degree to which the consumer views different products as substitutes is influenced by the functionality and quality of the offering, the attractiveness of the price of the substitute, and the cost the customer must incur to switch to substitute products. Organizations can reduce the threat of substitute products by differentiating their products and reinforcing in the minds of the customer the relative advantages of their offering relative to the substitute. For example, the design of eyeglasses has improved so much that they have become much more of a fashion accessory than in the past, and customers are choosing them simply because they look good. Lens manufacturers need to be well aware of developments in the laser surgery sector because it has the potential to wipe out their industry. Likewise, breweries need to be aware of not only the actions of producers of direct substitutes to beer, such as wine, but also the explosion of coffeehouses, such as Starbucks, that give consumers another place to socialize and unwind.

BARGAINING POWER OF SUPPLIERS

Organizations rely on suppliers to provide them with raw materials that they transform into outputs demanded by the customer. With the advent of approaches such as total quality management and supply chain management, the relationship between producer and supplier has improved significantly across many value chains. However, in sectors where there are only a few suppliers of specific raw materials, these supplier organizations can exert undue power within the relationship and negotiate better conditions. At the other end of the spectrum, suppliers who supply raw materials that are viewed as commodities have no power to influence the relationship and become price takers. The power of suppliers decreases relative to the number of suppliers serving the market, the ability of the buyer to replace their offering with substitute products, and the sector demand for the suppliers' output. If a small number of suppliers control access to scarce raw materials, they can determine the price and even refuse supply to new entrants, creating a barrier to entry.

EXAMPLE: One of the best examples of supplier power in recent years is that of Intel, which because of its dominant position as preferred supplier of microchips to the PC industry has been able to charge a premium price for its product. Its dominant position is based not only on the quality of its product but also on brand recognition on the part of the end customer. Producers such as Dell have been reluctant to switch to the rival supplier, Advanced Micro Devices, for fear that they would lose market share and have even gone as far as to allow Intel to place a label on the PC case stating "Intel Inside." However, recently a number of producers have begun to switch from Intel to Advanced Micro Devices' offering because of its perceived performance and cost advantages.

POWER OF THE BUYER

The influence of the buyer on the competitive forces in an industry is affected by many of the same factors that influence the bargaining power of suppliers. In industry sectors where there are few buyers relative to the number of suppliers or where one buyer represents a significant share of the market, these buyers can exert significant influence on the supplier–buyer relationship.

EXAMPLE: Taking the retail sector as an example, organizations such as Wal-Mart and Tesco can exert significant power across the sector because of their buying dominance. The power of the buyer is also stronger in sectors where suppliers are small and supply generic products that are readily substituted. The power of the buyer also increases if the costs of switching to alternative suppliers are low. Buyers in sectors that have been dominated by supplier power have increased their power by engaging in backward integration within the supply chain to create their own brand as an alternative to supplier dominance. Similarly, suppliers can forward integrate if buyer power becomes excessive in the sector (Schilling, 2006).

By analyzing the various forces influencing the competitiveness of its industry, an organization can identify the potential risks and opportunities that may emerge in the future. The organization can also analyze other markets to determine how attractive and open they are to entry. With this knowledge, the organization can begin to determine the goals it should pursue to best position itself for the future. Ongoing analysis of industry sectors is beneficial because it also allows the organization to iteratively assess the suitability of the goals it is pursuing, given emerging trends.

Benchmarking

Benchmarking is another technique used in goal planning. Benchmarking is the process of evaluating aspects of business processes in relation to best practice elsewhere. It typically involves visiting other organizations renowned for excellence in one or more of their processes. Competitive benchmarking involves looking at competitors' products, processes, and services to identify potential improvements that can add value to customers. Collaborative benchmarking involves noncompetitive organizations comparing their processes with each other. The more open the access to an organization's processes, the more valuable the output of the benchmarking process. The output of a particular benchmarking exercise is a list of requirements or opportunities that can improve the organization's competitive position. When choosing the organization to benchmark against, one often uses a checklist to score the prospective applicants. One example of a checklist is shown in Table 4.1.

This checklist has been divided into four areas that look at finance, operations, organization, and marketing. These criteria enable management to identify benchmarks of excellence in different domains that they may be able to study and imitate their success. The knowledge acquired during the benchmarking exercise becomes a powerful driver in developing goals and actions for the future.

SWOT Analysis

SWOT analysis is a popular technique used for environment analysis at the organizational level. The SWOT technique identifies the strengths, weaknesses, and opportunities of and threats to the organization. The strengths of the organization are the things that distinguish it from its competitors and must be maintained in the future. If these strengths have been developed by the organization over a sustained period and provide a real advantage over competitors, then they are sometimes called core competencies

Table 4.1   Sample Benchmarking Analysis Checklist

Criteria	Company A	Company B
Finance		
Return on Equity	34	45
Sales per Employee	23	34
Operating Cash Flow	1.76	1.55
Operations		
Labor Productivity	67	78
Material Use	78	90
Machine Use	67	68
Quality Throughput	99	87
Organization		
Absenteeism	2	3
Attrition Rate	4	4
Staff-Manager Ratio	20	25
Direct-Indirect Ratio	300	410
Marketing		
Market Share	34	45
Advertising Costs	345	456
Retention Rates	45	100

(Hamel & Prahalad, 1990). The weaknesses are the things that need improvement and are often identified through benchmarking against other organizations. In this way the organization learns about the state of the art and their relative capability. The strengths and weaknesses are the current state of the organization in the industry in which it competes. Opportunities and threats describe the future of the organization. Opportunities are the things that may happen in the future and that the organization can exploit for benefit. Threats are the things that have the potential to damage the organization in the future. SWOT typically has internal and external perspectives. Employees and managers provide internal perspectives on the strengths and weaknesses of an organization. Customers and other stakeholders can provide an external perspective.

STRENGTHS

These are statements of the strengths of the organization. Typical questions that may need to be asked include "What advantages does the organization have?" "What does the organization do better than similar organizations?" "What unique resources does the organization have access to?" and "What do stakeholders identify as strengths?" Strengths need to be gauged in relation to stakeholders' expectations and the strengths of competitors. If stakeholders expect high quality and competitors deliver high quality, then delivering high quality is not a strength. Examples of strengths include customer loyalty, reputation, cost leadership, skilled labor, technological capability, and infrastructure. Improving strengths keeps organizations ahead of their competitors in those specific areas.

WEAKNESSES

Typical questions that may need to be asked include "What needs to be improved?" "What needs to be avoided?" and "What do stakeholders identify as weaknesses?" Unpleasant truths may need to be faced. Benchmarking may identify strengths in other organizations that may translate into internal weaknesses. Examples of weaknesses include poor management, incorrect strategic direction, inefficient labor, obsolete processes, and technology. Weaknesses identify where the organization has room for improvement relative to the environment in which it competes.

OPPORTUNITIES

External opportunities arise from changes such as shifts in technology, government policy and regulations, social behavior of customers, and competitive behavior in the sector. A useful way to look for opportunities is to examine strengths and weaknesses in detail. Both strengths and weaknesses can provide new opportunities. Examples of opportunities include the possibility of competitors leaving the market, access to low-cost raw materials, expanding market, and strong knowledge of technological platforms. Whereas weaknesses exist in the present, opportunities exist in the future.

THREATS

Threats are future potential events that may weaken the organization's position in the market. Threats can originate internally or can be created externally. Internal threats can include overreliance on certain employees who may move to a competitor or on a potentially unreliable technology. External threats may come from competitors or the broader external environment. Identifying threats may put existing problems into perspective and allow the organization to plan to avoid these pitfalls. Examples of threats include entry of new competitors, poor supplier relations, and technology from other industries disrupting the sector.

Core Competencies

Core competencies can be viewed as the integrated set of abilities that distinguish an organization in the marketplace (Schilling, 2006). These competencies usually are focused on specific products or processes and provide the organization with its strategic advantage. Usually an organization has no more than five or six core competencies. A core competency (e.g., logistics) often combines a number of key skills, which makes it difficult for other organizations to replicate. This offers an organization an opportunity for innovation because it can perform these core competencies better than competitors and thus can add value to the customer. By understanding what the core competencies are, the organization is able to focus on strategic business development in key areas where it has a real advantage over the competition. However, core competencies depend on the ability to deliver value to the external environment. Because this environment is in a constant state of flux, core competencies can become obsolete. Managers must periodically assess the relevance of their core competencies to the external environment. If they don't, then they will focus on innovative developments in core areas that may no longer add value and will not achieve the desired result. The identification of core competencies can also be detrimental to an organization's innovation effort if it is allowed to limit the avenues employees explore for opportunity.

Because of the turbulence of the modern environment, many organizations are trying to develop dynamic capabilities that allow them to adapt to new opportunities faster than their competitors. Some core competencies embody the organization's ability to innovate. These core competencies include agility, flexibility, and adaptability. Core competencies can also include an organization's ability to understand market needs, form strategic partnerships, or manage projects, all of which support the innovation process. Innovation and innovation management are competencies that every organization strives to make core competencies.

Developing Statements

The purpose of statements is to communicate key messages and beliefs across the organization that will guide the behavior of all members. Statements need not exist only at a corporate level but can be cascaded down to individual units and departments. The presence of statements clarifies and communicates focus and highlights conflicts. Statements should be clear and concise to facilitate communication and understanding of direction across the organization. Long and overly complex statements can be difficult to understand. Such statements often remain as documents in a strategic manual rather than a motivating force to engage the organization. The two statements most commonly developed by an organization are the mission and vision statements. Other statements include quality, customer, and safety statements.

MISSION

This is a statement of the organization's purpose. The mission statement should be a succinct representation of the organization's current purpose or reason for being. It often incorporates meaningful and measurable criteria. The mission statement is broad enough that it can reflect the basic principles of the organization over the decades and acts as a guide to sustaining the core beliefs of the organization. The mission statement is a key consideration for anyone evaluating the strategic direction of the organization.

VISION

This is a statement of what the organization needs to achieve. It is a statement of what the organization wants to be in the future. A vision of the future is something that is credible and realistic but significantly better than what exists today. The vision is a powerful motivational tool because it communicates where the organization is moving. A common metaphor for organizational vision is the image of a lighthouse that guides a ship during a storm. At any point during the journey, the crew can readjust course, if necessary, or decide that their destination is no longer appropriate. Without the lighthouse, they would be lost at sea, without any points of reference to go forward or back. The vision statement acts as a guide both for the other goals and for the generation of actions to move the organization forward.

SAFETY

This is a statement of the safety policy of the organization. Safety statements have become popular in recent years to protect the health and safety of employees and customers.

QUALITY

This is a statement of the quality policy of the organization, that is, the goals of the organization with respect to quality assurance and quality control of its products, processes, and services.

Many other statements can be defined for the organization and include statements of core values, core competencies, and critical success factors. They all contribute to communicating the values, beliefs, and direction of the organization and provide a guiding influence for defining the associated goals of the organization that will in turn influence the innovative actions it undertakes.

EXAMPLE: Sun Hotel is a medium-sized three-star family-run hotel with 150 employees. Their mission is to offer accommodation to the business traveler who values culture and family values. Their vision is to grow business internationally through alliances with similar culture-focused hotels and expand into the tourist market. SWOT analysis revealed the following statements.

Strengths

- Family tradition
- Many years of experience and culture of hospitality among staff
- Multicultural staff

Weaknesses

- Old infrastructure
- Weak marketing
- Aging (retiring) management
- Lack of experience in international cooperation

Opportunities

- City plans new annual festival to increase tourist market.
- City has great tourist potential.
- Government is expected to launch business expansion tax incentives.

Threats

- New hotels and hotel chains in city
- More demanding business travelers
- Health and safety legislation

Summary

The process of defining goals requires organizations to consider creating statements of mission and vision, stakeholder requirements, strategic objectives, and performance indicators. A number of tools can be used to analyze the organization's environment and inform such a process, including benchmarking, PEST analysis, and SWOT. A statement of, for example, the organization's vision is one of the first steps in generating a comprehensive set of goals. In the next chapter we continue this process of developing goals by looking at how to determine the stakeholder requirements of the organization and subsequently develop appropriate strategic objectives that move the organization forward through the innovation process.

Activities

This activity requires you to create a number of statements for your organization, including mission, vision, safety, and quality statements. Undertake an analysis of your chosen organization. As part of this analysis, define a concise list of your organization's strengths, weaknesses, opportunities, and threats. Copy Table 4.2 into a spreadsheet and complete the fields. Place the name of the statement in the "Group" column and the statement detail in the "Title" column. Create as many rows as needed for your statements.

STRETCH: This activity may include other elements such as creating a separate list of PEST statements or statements of core competencies divided into appropriate headings. Assess whether these new statements should result in revisiting the definition of the vision statement developed earlier.

REFLECTIONS

- Outline the process of defining goals for an organization.
- Explain the benchmarking process.
- Explain the difference between rational and incremental goal definition.
- What do each of the letters in *PEST* stand for?
- How do strategic objectives differ from performance indicators?
- Define a mission statement.
- How does mission differ from vision?

Table 4.2　　Create Statements

Statements		
Group	**Title**	**Status**
Mission		
Vision		
Core Value		
Core Value		
Quality		
Safety		
Strengths		
Strengths		
Strengths		
Strengths		
Weaknesses		
Weaknesses		
Weaknesses		
Weaknesses		
Weaknesses		
Opportunities		
Opportunities		
Opportunities		
Threats		
Threats		
Threats		
Threats		
Threats		
Threats		
Threats		

Group: Label of the statement (e.g., "Strengths," "Weaknesses")
Title: The statement in less than 12 words
Status: Status of the requirement (e.g., "Not Started," "In Progress," "Waiting," "Completed")

Defining Objectives 5

Defining objectives provides a stimulus for the generation and implementation of ideas and other actions. By aligning innovative actions with the strategic objectives, the organization can move toward its vision of the future. Strategic objectives begin by understanding the demand from the main stakeholders of the organization. These stakeholders can come mainly from outside the organization. Understanding requirements, together with environmental analysis, provides organizations with insights into the future and the goals they should pursue to achieve maximum advantage. Common stakeholders in organizations include shareholders, customers, regulatory authorities, suppliers, and staff. Stakeholders place requirements on the organization that must be met if the organization is to be sustainable in the long term. Stakeholder requirements form a core part of organizational goals and act as a driver for innovation. For example, shareholders often demand more turnover and lower costs in order to maximize value. Customers often demand new products, better services, and faster responses. Regulatory authorities demand conformance to certain regulations (e.g., health and safety or environmental issues). The various strategic objectives that an organization selects determine how it will address these requirements in the future. Strategic objectives can be viewed as key decisions around which the organization will choose to focus its innovation.

LEARNING TARGETS

When you have completed this chapter you will be able to

- Describe the importance of stakeholders in the innovation process
- Explain the terms *transactional* and *contextual requirements of stakeholders*

- Define *strategic thrusts*
- Understand how to develop strategic objectives
- Explain the importance of concise strategic plans
- Discuss how to evaluate strategic objectives
- Understand the need to monitor emergent objectives

Identifying Stakeholders

Stakeholders are both individuals and other organizations that have a stake in the operation and success of the organization. They place demands on the organization. We naturally think of shareholders as stakeholders in a profit-making organization. For example, shareholders typically require higher returns for their continued engagement in the organization (e.g., profits). Customers are another common stakeholder. Customers choose products, processes, or services because they meet certain expectations and fulfill certain needs. Continuing to meet these expectations means continuously listening to customers' requirements such as more features, lower cost, or perhaps faster delivery. Other stakeholders are less apparent. Regulators are stakeholders who issue requirements on issues such as health and safety, waste emissions, or even taxation. They require that the organization conform to particular regulations and standards. Employees are another important stakeholder group. Employees regularly issue requirements such as greater job security or perhaps better prospects of promotion. Suppliers are also stakeholders in the organization and require that the organization continue to purchase products and services from them and do so at a satisfactory price. Stakeholders vary across organizations. Identifying the organization's stakeholders is a key step in defining innovation goals. Eight types of stakeholders are common (Shapiro, 2001) and are illustrated in Figure 5.1.

CUSTOMERS

These stakeholders consume and use a particular product, process, or service. They can be external customers such as patients in a hospital or consumers of television sets, or they can be internal customers within a larger organization who use output from another section. For example, doctors can be internal customers of a hospital computer service department, and managers can be internal customers of an organization's human resource department.

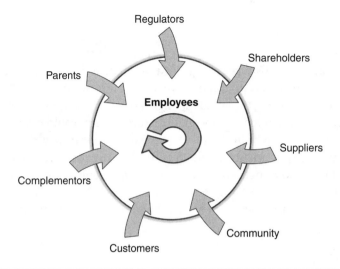

Figure 5.1   External and Internal Stakeholders

SUPPLIERS

These stakeholders are suppliers of products or services and even processes to the organization. Like the other stakeholder groups, they may have requirements that can include maintaining a stable and effective supply chain. Typical supplier requirements include payment on time, less bureaucracy when delivering goods, or more accurate information. With the advent of concepts such as the extended enterprise and supply chain management, suppliers have been viewed more as strategic partners of the organization rather than simply inputs to the process. As their requirements are addressed, relationships between the organizations can grow and increased efficiency can be attained. Suppliers not only influence the goals pursued by the organization but are also valuable stimuli for the action process. Suppliers can often introduce innovative technology to the customer organization. Fulfilling supplier requirements can enhance the innovative capability of the organization.

SHAREHOLDERS

These are individuals and other organizations that have invested in and own a share of the organization. For example, a venture capitalist who has invested funds in the organization will want the organization to achieve its targeted market share, keep expenses low, and ultimately maximize the return on his or her investment.

PARENTS

These are parent organizations. For example, a manufacturing plant in Ireland may be a subsidiary of a parent organization in the United States. The parent organization imposes requirements on the manufacturing organization such as lower costs, higher quality, and lower inventory. A quality department in the same manufacturing plant would have the senior management team within the plant as a parent whom they report to and who impose requirements on the quality department. It is interesting to note that the requirements of parents are often articulated through a parent innovation plan. In this way goals can be linked between organizations through the hierarchical layers within the same macro organization.

REGULATORS

These government and other regulatory entities have some degree of influence over the business. Most organizations need to conform to financial regulations, health and safety regulations, and so on. Regulations can also include internal policies and standards developed inside the organization. The quality department in a large organization can issue internal regulations to, say, production departments regarding production practice. Regulators normally impose requirements on organizations for the benefit of the wider community. Regulations, such as those imposed by the Food and Drug Administration on the pharmaceutical sector, may be viewed as a hindrance to the organization, but they are designed to prevent potentially harmful drugs from making their way into the public domain. Regulators ensure that the organization maintains a long-term perspective that benefits communities by imposing regulatory barriers on trade.

COMPLEMENTORS

These are partners and other groups that add value to the overall product, process, or service such as strategic partners, joint ownership partners, and some key suppliers, distributors, and customers. Other organizations depend on the organization's product for their continued success (and vice versa) and so need to at least be aware of the organization's innovative activities. Some of the factors that influence the impact of complementors on the organization are the importance of complementors, the number of alternative complementors available, and the party that captures the value generated by the complementarity between offerings (Schilling, 2006). If firms can improve the synergy between their products and those of their complementor, then they can strengthen their competitive position in the market and create barriers to entry. The risks in this relationship are that one party will become dominant over the other and revert to power over the supplier or buyer or may decide to reverse or forward integrate in the business to improve their competitive position.

EXAMPLE: An example of complementors is the relationship between the Sony PlayStation and game designers such as EA Games. The presence of a wide range of games on the PlayStation platform increases sales of the console unit. If a major game designer decided for some reason not to support a particular console maker, then this would act as a significant barrier to that console's success in the industry.

INTERMEDIARIES

These are consultants, quasi customers, and other specialist groups that assist in product development and service provision. For example, the retail outlets for the organization's product may have requirements such as more robust packaging or longer shelf life. By meeting these requirements, the organization not only enhances the operations of the retailer but also enhances the market perception of their own product because less non-conforming product will reach the end customer.

EMPLOYEES

Employees can also be key stakeholders in any organization and will have requirements that they need the organization to fulfill. These requirements can include better job security, higher pay, greater empowerment, or more opportunity for development. If the employee requirement is linked to a certain strategic objective, then employees will be motivated to develop appropriate innovations to achieve these goals.

COMMUNITY

This is the local, regional, and national community in which the organization exists. Community stakeholders can also include potential partners in the future. Developments such as open innovation (Chesbrough, 2003) and lead users (von Hippel, 1994) illustrate that ever more diverse people are placing requirements on the organization. Understanding market dynamics, technological forecasting, communicating and working with users, and listening to the voice of the customer are ways in which these stakeholder requirements may be defined. These activities allow insights into how the organization can best develop in the future. Determining needs can be a challenging task because people often are unable to recognize their needs. Identifying the hidden needs

(Goffin & Mitchell, 2005) of the various stakeholders can be crucial to meeting the future needs of the organization.

Defining Requirements

Understanding the stakeholders in an organization, together with environmental analysis, is a key step in defining appropriate innovation goals. Discovering what each stakeholder requires can be a difficult task, however. Stakeholders can have a transactional or contextual relationship with the organization (Pava, 1983). A transactional relationship exists where the organization can influence and change the stakeholder's requirement in some way. Influencing a shareholder to change his or her expectation from higher profits to greater investment is an example of a transactional relationship. A contextual relationship exists where the organization has no influence on the requirement and where a stakeholder insists on demands being met. For example, corporate owners may insist on a particular cost saving target or increase in revenue. Government regulators are another example of a stakeholder with a contextual relationship.

Requirement statements are expressions of demand from stakeholders. Some stakeholders may have many requirements, whereas others may have only a few. Many requirements may need to be distilled and summarized into the key few. Many techniques are used for defining requirements. One simple technique is to ask each stakeholder to articulate what he or she perceives as the weaknesses of and threats to the organization's offering. These can be listed and grouped into a number of categories. When wording the requirement, it is useful to use language expressed from the perspective of the stakeholder. For example, the customer would express a requirement such as "deliver products faster" or "provide more reliable products." Put yourself in the position of the stakeholder and use his or her voice to express the requirement. The use of an active verb in the description can also be useful (e.g., *deliver* and *provide*). More structured techniques for gathering requirements include objective data analysis, Delphi forecasting, focus groups, and conjoint analysis.

OBJECTIVE DATA ANALYSIS

This is a technique in which requirements are derived through the analysis of large amounts of data. Data such as historical sales, behavior of machines and processes, previous performance measurements, and customer complaint logs are used to predict the system's behavior in the future. Predicted poor performance can be translated into requirements that, if addressed, will result in the system behaving better.

DELPHI FORECASTING

This is a qualitative technique for predicting future requirements of stakeholders. It is based on human judgment rather than the analysis of data. The Delphi method replaces direct open debate, in which one personality may dominate others, with a carefully designed program of interrogation. Three principal techniques are used: anonymous interaction, iteration with controlled feedback, and statistical group response.

FOCUS GROUPS

This is a qualitative approach in which a group of 8 to 12 people are brought together to discuss a common theme. The focus group normally is led by a moderator who directs the discussion though questioning. Common groupings include existing and potential customers, suppliers, and lead users. The benefit of a focus group is that discussion can be freer flowing than in a one-to-one situation, and the comments of one person can illuminate the discussion of another. An important aspect of a focus group is a well-trained moderator, because the moderator must ensure that the discussion stays focused on the desired theme or concept. Focus groups are valuable in that they can identify needs and desires of potential stakeholders in a speedy and economical manner. Focus groups are often used as part of the market research on a product innovation in order to identify the features and attributes that the eventual customer will value.

CONJOINT ANALYSIS

One of the most useful techniques to emerge in recent years to help organizations prioritize prospective innovation attributes is conjoint analysis. Conjoint analysis is a statistical technique that an organization can use to determine what combination of a limited number of attributes is preferred by its customers. This stated preference model works by first creating a definite list of attributes that will be tested against sample customers in order to establish the relative utility level of each. Customers are interviewed and asked to choose between two versions of the product, each with certain attributes. By repeating the exercise a number of times and analyzing a customer's tradeoffs with the various product versions, the researcher is able to calculate the relative preference of attributes and the tradeoffs that the customer makes. Key challenges associated with use of the technique include defining the correct attributes to be tested and identifying a sample of respondents that adequately reflects the true population.

EXAMPLE: Clearview Pharmaceuticals is a small startup manufacturing company. The innovation team is responsible for mainly process innovations and includes key personnel from all of the main functions in the organization—all managers and some specialists. They meet bimonthly to discuss the progress of their goals and review the status of various projects. They also review any ideas that have been generated by employees that match organization goals. The requirements of each stakeholder have been determined and are monitored regularly by specific members of the team (Table 5.1). A key requirement of customers is reducing manufacturing lead times. At each meeting the production manager gives an update to the rest of the team on the status of this requirement. As new requirements are determined, they are added and responsibility is assigned.

Activities

This activity requires you to create a list of 10 to 15 stakeholder requirements for your organization divided among your key stakeholders.

Table 5.1   Clearview Pharmaceuticals Stakeholder Requirements

Requirements			
Group	**Title**	**Responsible**	**Status**
Conformance	Maintain FDA Conformance	Mike Mannion	☺
Conformance	Achieve ISO 14001 certification	Mark Ryan	☹
Conformance	Achieve ISO 9001 certification	David Jones	☺
Parent	Increase productivity	David Jones	☺
Parent	Modernize information technology infrastructure	Andy Scott	☺
Parent	Lower purchasing costs per unit	Mark Ryan	☺
Customer	Reduce product costs	Mary Joyce	☺
Customer	Reduce delivery time	Paul Jones	☹
Customer	Online services	Mike Mannion	☺
Shareholder	Return on profit	Mike Mannion	☺

Identify the names of the stakeholders for your organization and list them in the "Group" column in Table 5.2. Identify at least two requirements for each stakeholder and list the detail of the requirements in the "Title" column. When articulating your requirements, try to use an active verb in the sentence, preferably at the beginning, and keep the number of words to a minimum (e.g., a supplier might require that your organization "improve the access to master schedule information"). Try to also use the voice of the stakeholder. Put yourself in the stakeholder's shoes and try to imagine what he or she may demand from your organization in his or her words. Assign responsibility for each of the requirements to a suitable employee of your organization. Assign a fictitious status to the various requirements to communicate how well they are being fulfilled by the organization. Copy Table 5.2 into a spreadsheet and complete the fields.

Stretch: Set up a focus group with other students in your class. Tell them about your product or service and ask them to identify a number of requirements that they feel they could realistically impose on your organization if they were customers of the product or service.

Strategic Plan

The strategic plan can be broken down into two related areas: strategic thrusts and strategic objectives. The strategic thrust identifies a specific area of focus. Different industry sectors will place greater focus on particular thrusts, given their importance to the industry. For example, an electronics organization will place high priority on technology as a strategic thrust, given its importance to survival in their sector. The term *strategic objective* refers to the more specific objective that the organization wants to achieve within the area of the strategic thrust. The term *strategic objective* relates to the choices that the organization makes in order to develop. The following are high-level strategic thrusts popular in many organizations (Hayes et al., 1988): capacity, facilities, technology, vertical integration, workforce, quality, planning, and organization.

CAPACITY

This thrust deals with issues of increasing, decreasing, or maintaining capacity. *Capacity* refers to machines, labor, and facilities. Typical choices within this thrust include demand management, outsourcing, floor space use, job enrichment, multiskilling, and work cells.

Table 5.2   Create Requirements

Requirements			
Group	**Title**	**Responsible**	**Status**

Group: Name of the stakeholder (e.g., "Customer")
Title: Title of the requirement (use the voice of the stakeholder, e.g., "Shorter lead times")
Responsible: Individual responsible for reporting the status of the requirement
Status: Status of the requirement (e.g., not started, in progress, waiting, completed)

FACILITIES

This thrust deals with the facilities that are used by the organization in meeting customer requirements. Facilities include issues related to location, floor space, plant facilities, and machine facilities. Choices regarding specific strategic objectives the organization may want to achieve in this area include capping the size of plants, developing focused lines and processes, and moving to an alternative location.

TECHNOLOGY

This thrust deals with decisions about the technology used in the organization such as technological platforms used in products and services, machinery and computer networks, and telephone exchanges. Choices can include issues related to appropriate standards and platforms to adopt, networking replacement of obsolete technology, and improving the communication infrastructure.

VERTICAL INTEGRATION

This thrust relates to the integration of the organization with suppliers and other strategic partners such as distributors, customers, and other stakeholders regarding finance, processes, technology, location, sharing of information, co-design, and so on.

WORKFORCE

This thrust deals with issues related to the human capital in the organization. Strategic choices that an organization might make in this area include improving delegation, improving reward and recognition, reducing bureaucracy, and developing knowledge workers.

QUALITY

This area attracts significant attention from service- and process-based organizations. Strategic choices in this area can include organizations setting objectives regarding enhanced products, processes, and services, reducing customer complaints, and improving environmental and health and safety compliance.

PLANNING

This thrust deals with strategic choices regarding material planning, order flow planning, product design planning, shop floor control, logistics, supply chain management, and so on.

ORGANIZATION

This thrust relates to choices regarding objectives such as alterations to management structures, control mechanisms, interaction between functional departments, management systems, and communication infrastructure.

Organizations choose to adopt the strategic thrusts most appropriate to move themselves forward into the future. In certain instances organizations select strategic thrusts based on well-known methods or frameworks. One such method is the balanced scorecard (Kaplan & Norton, 1996), which identifies four major thrusts: finance, customers, internal processes, and learning and growth. Other methods that provide strategic thrusts include ISO 9000, the European Foundation for Quality Management model, and the Malcolm Baldrige Award criteria. Even when an organization chooses to use a well-known method, it cannot be emphasized enough that each organization will need to identify strategic thrusts that are most appropriate to that organization and then decide the specific strategic objectives needed to bring the organization into the future. The information gathered as part of the organization's environmental analysis and stakeholder requirements will drive many of the strategic choices that the organization makes. If the organization achieves its strategic objectives over time, then it will have addressed the demands identified in the planning phase and strengthened its competitive position. Although certain organizations may share common strategic thrusts, the more specific strategic objectives will be unique to the individual organization.

Generic Objectives

An organization can choose from a number of different types of strategic objectives to pursue. These types provide the organization with a direction and focus from which to generate supporting innovative actions. Porter's (1980) research identified three generic strategies that an organization may want to pursue in order to attain enhanced competitive advantage: cost leadership objectives, differentiation objectives, and focused (niche) objectives.

COST LEADERSHIP OBJECTIVES

Organizations that adopt this strategy strive to be the industry leader in terms of cost. These organizations often are not the cheapest in the market, but by increasing efficiency they can make a higher return than their competitors. Strategic objectives under this generic strategy seek to improve efficiency and control costs throughout the organization's supply chain, achieve economies of scale, improve experience and learning, and reduce product complexity and variety. Organizations compete with each other in areas such as process technology, raw material costs, and capacity use and usually sell a standard product. Examples of organizations pursuing this strategy include low-cost airlines such as Ryanair and Southwest Airlines. Although these organizations would define cost leadership as their strategic objective, they would have a number of sub-objectives that are specific to the organization's environment and core competencies. Examples of these include "short-haul, point-to-point routes, often to secondary airports," "standardized aircraft fleet," and "limited passenger services."

DIFFERENTIATION OBJECTIVES

Differentiation occurs when the product or service provided by the organization meets the needs of some customers better than those of its competitors. Underlying differentiation is the concept of market segmentation and the knowledge that specific groups of customers have unique needs. Organizations that focus on this category of objective seek to add value to specific market segments by differentiating both the tangible and intangible features of their products and services. Strategic objectives under this heading typically involve applying technological superiority over competitors, outperforming competitors in an area of quality, and providing better support services or lead time to the customer. This category is populated by organizations that strive to be unique in the industry and stand out from competitors. Although organizations pursuing a particular differentiation strategy can gain a competitive advantage over their competitors, the risks associated with such a strategy are that it will introduce costly features not valued by the customer, or if the organization is successful, then competitors will copy the differentiated product and thereby reduce the market share. Volvo is an example of an organization that has adopted a differentiation strategy that has driven innovation and success. Their commitment to safety in all their market offerings has encouraged customer loyalty and increased sales. Although other competitors have followed Volvo's lead, customers still perceive the company's product as a safer car.

FOCUSED (NICHE) OBJECTIVES

A focused strategy occurs when an organization focuses on a specific niche in the marketplace. The organization attains its competitive advantage by meeting the unique needs of this niche market better than any other organization. Leadership can be achieved by adopting cost leadership or differentiation strategies, which are designed specifically for the environment of the niche market. This category is populated by organizations that strive to become leaders in a specific market segment. The risk associated with this strategy is that the segment will be too successful and draw other organizations into the market niche. Another risk is that the market niche will be too small to sustain the organization over the long term.

Strategic Objectives

Strategic objectives are the actual objectives, strategies, or choices that an organization strives to implement over a planning horizon (typically 1–5 years) (Mintzberg, Quinn, & James, 1988). An important consideration when defining and choosing objectives is the ability to implement them. Strategic objectives must be chosen that have a chance of being implemented with the available resources—time, people, and money—over the planning horizon. Attempting the impossible can be bad for morale and can lead to greater resistance to change in the future. There must be a balance between the objectives set by the organization and the resources and capabilities available to achieve these objectives. Strategic objectives should be neither too general nor too specific. Objectives that are too general fail to give guidance to people during the idea generation process. Objectives that are too specific take away power from the people generating ideas and tend to be more akin to projects. Strategic objectives should inform and stimulate the idea generation process of the organization so that people can develop ideas and projects at an operational level that helps the organization achieve those objectives. Objectives that are clear and transparent allow employees to translate them easily into projects that can move the organization toward its goals. Clear objectives also facilitate better decision making when one is aligning projects with objectives.

Although analysis of the external environment and determination of the appropriate strategic thrusts are important to an organization, it is the specific strategic objectives that determine where the organization wants to go and how it is going to get there. The selection of specific strategic objectives depends on a number of internal and external factors. The internal factors can include the current technical and organizational capabilities, the success of the current business model, the available resources, and the organization's vision statement for the future. External factors that may shape an organization's strategic objectives include external network capabilities, industry structure, competition, and the rate of technological change (Davila, Epstein, & Shelton, 2006). Strategic objectives such as consolidation, market penetration, product and market development, and diversification are just some of the options open to an organization. Although these are based on the organization's unique context, their appropriateness should be reviewed in terms of suitability, acceptability, and feasibility (Johnson & Scholes, 2002).

SUITABILITY

This is the fit between the strategic objective and the circumstances in which the organization is operating. One of the key questions an organization must ask is, "Does this objective make sense given this organization's competitive position?" Objectives that are not suitable for the organization's context are unlikely to result in their desired outcome or move the organization in the direction it needs to go.

ACCEPTABILITY

This is concerned with the expected outcomes of the strategic objective in terms of performance and its alignment with the organization's expectations. Acceptability of an objective can be viewed as a synthesis of three elements: return, risk, and stakeholder reaction. Return is the benefit that will accrue to the organization from pursuing a specific objective and the value of the return relative to the required inputs. Risk is the potential drawbacks of the objective. Risk can relate to the return from pursuing the objective but can also relate to other characteristics of value, such as threatening the organization's mission or culture. If the risk associated with an objective is high, then it might be unacceptable to the organization. The last element influencing the acceptability of a particular strategic objective is stakeholder reaction. As discussed earlier, various stakeholders have different vested interests in the organization. If a proposed objective clashes with the values or requirements of particular stakeholder groups, then this can lead to potential conflict and resistance. Organizations must reflect on the impact of particular strategic objectives on powerful stakeholders and ensure that conflict is minimized.

FEASIBILITY

This is concerned with the organization's ability to implement a particular objective. Because organizations have different competencies, skills, experience, and financial resources, certain organizations are more likely to achieve certain strategic objectives than others. Ensuring a good match between the objectives pursued and the organization's capabilities can provide a competitive advantage in implementing these objectives. When organizations pursue objectives that are not aligned with their capabilities, the feasibility of implementing these objectives successfully diminishes and the associated risk levels increase.

Screening the various strategic alternatives, an organization can identify the most suitable objectives that it can pursue under each of its strategic thrusts. The organization further defines these objectives by linking them with associated performance indicators. High-level organizational goals become more tangible and understood across the organization and enable people to contribute potential innovations that can help the organization achieve its objectives, fulfill stakeholder requirements, and ultimately achieve its vision for the future.

Objectives for Innovation

There are many potential objectives for any organization. Many of these objectives will lead to changes to the organization's products, processes, and services. Reflecting back on the difference between operations and innovation, most objectives refer to changes in operations or in how things work at present (products, processes, and services). However, the innovation activity itself can also be subjected to change. The primary things that can be improved with this activity are the process itself (i.e., how the process translates ideas into innovations) and the resources it uses (i.e., tools, techniques, people, and funding). This book is about describing an effective innovation management process and identifying some of the tools and techniques used to resource it. Goals can be defined for guiding innovations to products, processes, and services that will flow through the innovation process, but goals can also be defined that focus on increasing the efficiency and effectiveness of the innovation process itself. Therefore, organizational goals can have a dual influence: on the operational reality of the organization and on the process by which the organization will transform itself from its current to its future position.

EXAMPLE: SwitchIt is a design and manufacturing company of electrical switch gear. All managers form a team that reviews their innovation plan weekly. The team has a set of strategic objectives for development over the period of 3 years. Nine strategic groups or thrusts have been defined, with a number of objectives or strategies per group. A sample of the strategic plan, showing only three of the thrusts and nine of the objectives, is shown in Table 5.3.

Table 5.3 Strategic Objectives at Switchlt (in Part)

Objectives			
Group	**Title**	**Responsible**	**Status**
Capacity	Use low-risk strategy for capacity expansion	Mary Roche	☺
Capacity	Improve capacity analysis techniques	David Noone	☹
Capacity	Improve labor flexibility for capacity changes	Michael Clark	☺
Capacity	Explore make-vs.-buy opportunities	Stewart O'Neill	☺
Responsiveness	Collaborate on development of more accurate forecasts	Danny Mulryn	☺
Responsiveness	Explore manufacture-to-order processes	Michael Clark	☺
Responsiveness	Reduce order delivery times	Stewart O'Neill	☺
Responsiveness	Improve dealer and supplier partnerships	Stewart O'Neill	☹
Organization	Migrate toward flatter and leaner organization	Danny Mulryn	☺

Summary

Stakeholder requirements place demands for innovation on the organization. Many of these demands lead directly to new ideas. Other requirements inform the strategic objectives and performance indicators of the organization. Some stakeholders are internal, but most are external. Some external stakeholders can have their requirements influenced by the organization, whereas others cannot. Requirements are ultimately translated into strategic objectives and performance indicators in the innovation management process. A strategic plan typically contains a large number of objectives divided into groups called strategic thrusts. A strategic plan is a living document. After it has been created, it can be edited and updated from time to time to reflect shifts in the external environment. The organizational efforts toward its objectives must be monitored continuously. This chapter has shown how to create a concise plan for any organization. Strategic plans are a guide for the innovative actions that the organization will pursue and must allow some room for interpretation, or they will constrain the type of innovation. The next chapter examines how organizations develop performance indicators to measure their progress toward the objectives they have defined.

Activities

This activity requires you to create a list of 16 to 24 strategic objectives for your organization, divided into strategic thrusts. First, define a minimum of three strategic thrusts and list them in the "Group" column of Table 5.4. The thrusts selected should be based on the environmental analysis and statements you defined earlier. Once you have created appropriate strategic thrusts, define at least two objectives for each and place them in the "Title" column. When articulating your objectives, try to use an active verb in the sentence, preferably at the beginning, and keep the number of words to a minimum. Try to define strategic objectives that are general enough to remain relevant for the entire planning period (e.g., 3 years). Avoid strategic objectives that can be achieved in 6 months or less; these may be defined more accurately as projects later. Once you have defined your strategic objectives, assign responsibility for their achievement to one or more members of the organization. Assign a fictitious status to the various objectives to communicate how well they are being fulfilled by the organization. Copy Table 5.4 into a spreadsheet and complete the fields.

STRETCH: Other elements in this activity may include discovering emergent objectives for organizations in your domain and creating a separate list titled "Emergent Objectives." These emergent objectives can be found in various trade magazines, online discussion forums, benchmarking, and so on and reflect general changes occurring across many organizations. These emergent objectives may lead you to revisit the strategic plan you defined earlier.

Table 5.4   Create Objectives

Objectives			
Group	**Title**	**Responsible**	**Status**

Group: Label of the strategic thrust (e.g., "Technology")

Title: Title of the objective (use an active verb, e.g., "Increase capacity in line with demand")

Responsible: Person responsible for reporting the status of the objective

Status: Status of the objective (e.g., "not started," "in progress," "waiting," "completed")

REFLECTIONS

- Define a stakeholder for a particular organization.
- Explain the terms *transactional* and *contextual requirements*.
- Define *strategic thrust.*
- List up to five strategic thrusts suitable for an organization such as a hospital.
- Explain why strategic plans should be concise.
- Discuss how to evaluate strategic objectives.
- Name four common or emergent objectives currently discussed in strategy literature.

Measuring Indicators 6

Measuring indicators is important to applying innovation because it highlights progress the organization is making toward its strategic objectives as a consequence of the innovative actions it has undertaken. This provides a feedback loop that allows the organization to adapt the innovative actions it will pursue in response to its success in achieving its performance targets and associated strategic objectives. Performance indicators are a measurable way of defining goals. Performance indicators set goals in the future that need to be met; they also monitor current progress toward those goals, and they provide a historical perspective on performance in the past. Indicators can be financial or nonfinancial. Financial indicators such as measures of revenues and cost have been popular in the past. An organization's worth is often measured by financial indicators such as turnover and profits. This is no longer solely the case, with many organizations having a value significantly greater than financial indicators suggest. This is because organizations are often measured in terms of their potential as knowledge organizations. A knowledge-based organization can change its products, processes, and services in response to changing market demands. Most performance indicators are moving toward nonfinancial indicators such as measuring absenteeism, rates of idea generation, and success of new products.

LEARNING TARGETS

When you have completed this chapter you will be able to

- Describe the difference between financial and nonfinancial indicators
- Understand the importance of indicators for motivating employees

- Explain the balanced scorecard technique
- List a number of current or emergent indicators used in many organizations
- Create a simple form for capturing the critical data for an indicator
- Indicate the key data points in a performance chart

Performance Indicators

Performance indicators are a measurable way of defining and monitoring goals. Performance indicators help to make goals tangible for the desired future state. They also monitor current progress toward future goals and provide a historical perspective on performance in the past. Indicators can be financial or nonfinancial. A framework of performance indicators that drive organizational goals down through organizational layers and motivate people to contribute to the achievement of these performance targets is essential to innovation.

An organization's worth historically has been measured by financial indicators such as turnover and profits. Although fiscal measures are an important indicator of overall company performance, they have been increasingly criticized for providing little indication of how that performance is achieved, how it can be improved, or how it may occur in the future (Neely, Adams, & Kennerley, 2002). Other criticisms have centered on the internal focus and historic perspective of financial measures, which promote short-term thinking and have little regard for customers. In the modern knowledge economy, organizational worth is strongly influenced by aspects such as knowledge and expertise, commitment of employees, and spending on research and development. In this respect, many of the performance indicators organizations now adopt are nonfinancial and focus on measuring issues such as absenteeism, rate of idea generation, and lead time for new product development.

Despite various corporate accounting scandals that have called into question the value of financial indicators, they do provide a partial and well-understood measure of the worth of an organization. However, knowledge-based organizations often have the potential to innovate and change financial performance based on underlying parameters that are not captured by the use of financial measures. The disadvantages of using financial indicators alone to reflect the progress of the organization include the following:

- Inconsistency with today's business realities
- Reliance on historical data ("driving by rearview mirror")

- Tendency to reinforce functional silos

- Sacrifice of long-term thinking

- Lack of relevance to many levels of the organization

- Lack of connection to strategic objectives

Many organizations choose to adopt a mixture of financial and nonfinancial indicators to determine the performance of the organization. Choosing nonfinancial indicators is advantageous because they are readily understood and engage individuals in contributing to their achievement. Financial indicators are often viewed as complicated, and people think they cannot influence their achievement as easily as certain nonfinancial metrics. The linking of defined performance indicators to the developed goals (in terms of statements, requirements, and strategies) can motivate people to generate initiatives that contribute to goal achievement. This is because they can relate to the performance indicators more easily than higher-level goals such as strategies and vision statements. The interlinking of the various levels of goals (statements, requirements, objectives, and indicators) drives the goals down through the layers of the organization (Kaplan & Norton, 1996) and ensures that efforts to achieve defined indicator targets contribute to fulfilling the strategic objectives of the organization and move it toward its vision.

Defining Indicators

Indicators show progress toward defined performance targets and motivate people to achieve goals. Key questions addressed by the performance indicator process include the following:

- What has happened in the organization?

- Why has it happened?

- Is the trend going to continue?

- What impact have efforts had on the trend?

Organizations that have defined performance indicators and manage their progress toward their targets are able to answer these questions and are better able to achieve their goals. Understanding indicators promotes idea generation across the organization and provides management with a feedback system to monitor the impact of their innovation effort. The choice of indicators the organization will pursue is contingent on their existing context and the goals the organization has set. However, all indicators can have the following attributes:

- The indicator is related directly to strategic objectives.
- The indicator is consistently repeatable over time, allowing comparisons.
- The indicator fosters improvement rather than monitoring.
- Measurements are reliable and verifiable.
- There is an appropriate mix of financial and nonfinancial metrics.
- There is a maximum number of measures.
- The indicator is simple and easy to use.
- The indicator provides fast feedback.
- Indicators can be linked in a hierarchy.

Performance indicators should be directly related to strategic objectives. This may not necessarily be a one-to-one relationship in that each objective has a unique indicator. It may be a many-to-many relationship in which any one objective or indicator can be related to a number of indicators or objectives, respectively. Indicators should be measurable over time; that is, the same indicator is capable of being measured easily over a long time period. This allows comparisons to be made between individual results or measures. If the trend is negative, then it can be used to generate ideas and projects. Indicators should foster improvement among the people who are measuring and reviewing the indicators. Indicators are meant to illustrate a trend that, if negative, will help to generate ideas that can reverse the trend. They are not meant to be used against the people who are responsible for carrying out the measurements. If this were the case, then these same people probably would find ways to manipulate the results in their favor. Indicators should be reliable and verifiable. Each indicator should measure a variable that is easily measured and offers direct comparisons to previous measurements. There needs to be an appropriate mix of financial and nonfinancial indicators (or, in many organizations, exclusively nonfinancial indicators). Within departments or project teams, most indicators may be nonfinancial, although in principle all indicators can be extrapolated and converted into cost at some level.

Opinions differ about the optimum number of indicators; conventional wisdom suggests that an organization should choose no more than seven indicators. From a practical perspective, fewer indicators can receive greater attention. Many organizations choose to measure a large number of performance indicators because they cannot decide which indicators are most important. The trouble is that too many indicators clutter up the decision-making process and communicate confusion. Indicators should be simple and easy to use and provide fast feedback. Indicators that take hours or days to measure are likely to be out of date and cause difficulties for the people responsible for their measurement. Indicators can also be linked hierarchically throughout a large organization. In this way an individual or team may be driven by only seven indicators, but the total number of indicators across a large organization can be significantly higher.

EXAMPLE: One of the first steps in choosing indicators is to look at macro indicators. All indicators stem from one of three macro indicators: cost, time, and accuracy. In recent years other macro indicators have been added to this list. They include flexibility, culture, and environment. An alternative way of using macro indicators is presented here under five headings:

Operations

Productivity (hours/unit)

Throughput (units per day)

Utilization (output/capacity)

Sales and Marketing

Sales per region

Sales per model

Marketing costs

People

Labor turnover

Overtime

Absenteeism

Research and Development

R&D expenditure

Failure rates

Additional revenue created

Environment

Emissions

Scrap and wastage

Accidents

Innovation Process Indicators

There are many potential performance indicators for any organization. Most of these indicators measure the impact that product, process, and service innovation has had on the organization's operations in terms of sales, market share, efficiency, speed, and so on. However, organizations can also choose indicators that measure the way they manage and execute innovation. These measurements often are subjective but provide a practical measure for assessing the organization's ongoing innovative capability. They also can act as a stimulus for generating ideas as to how the innovation process itself can be changed to improve its operational efficiency. Applicable indicators might include the following measures:

- Percentage of revenue attributable to recent innovations
- Percentage of ideas migrating to projects
- Number of projects per member of staff
- Percentage of staff involved in the generation of ideas or problems
- Percentage of actions originating outside the organization
- Percentage of indicators without actions
- Number of projects per strategic thrust
- Percentage of strategies without actions
- Percentage of actions delivered within planned constraints
- Percentage of actions abandoned during the innovation process
- Cost–benefit ratio of the portfolio undertaken

Each of these indicators provides a measure of the level of innovation in an organization and can provide important information about where the process can be improved. Through a suitable mix of performance indicators, an organization can focus on achieving its defined goals and enhancing the effectiveness of its innovation process.

Performance Charts

Every indicator must have a unit of measure. For example, productivity can be measured in terms of hours per unit or use in terms of the percentage of output over capacity. Each indicator consists of three major measurements: its origin, or what it measured at some point in the past; its current measurement; and its target performance, or what it should measure at some point in the future. Many organizations like to add the fourth measurement: its stretch target. The stretch target is a more ambitious target performance at some point in the future. It allows people to consider stretching beyond the current target to achieve breakthrough performance for the current period.

Ongoing trends of performance indicators over time can be represented graphically using a performance chart. The key attributes of the performance chart are illustrated in Figure 6.1. Each indicator has an origin date and an origin value. This is typically the beginning of a particular year. Each indicator also has a defined future target date and target value. Performance charts may also indicate the stretch target that may be associated with the particular indicator. Another characteristic of a performance chart is the record of values over the planning horizon (represented by the star symbol in Figure 6.1). Charts should be kept simple. As measurements are recorded on the chart, they reveal the progress that the organization is making toward its goals. In Figure 6.1, the organization is focusing on a percentage indicator such as yield, deliveries on time, or perhaps even student attendance. The organization's starting point at the beginning of 2006 shows that at this time it was achieving an 80% result. At this time it set itself the target of achieving 90% by the end of the year. It also set itself the more ambitious stretch target of 97% to achieve if possible within the same planning horizon. Although the organization started at a baseline of 80%, by the end of the year it had achieved its target and had made progress toward its more ambitious goal of 97%. This information is a powerful motivator for employees because they can see not only what is expected of them but also how their efforts affect particular indicators.

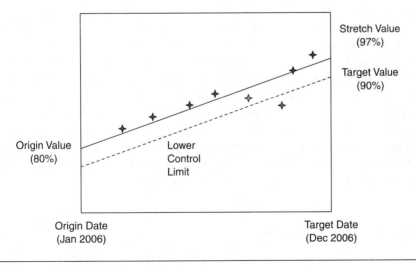

Figure 6.1   Performance Chart

The innovative actions undertaken in response to the defined performance indicators and associated goals will affect overall organization performance. The innovation process may deliver results such as higher sales and profits, lower costs, larger market share, more product families, and new markets. The achievement of these results moves the organization toward its desired goals and provides a feedback loop regarding the effectiveness of the actions undertaken. This information flow allows the organization to change the relative priority of individual goals across the planning horizon. As the organization's innovative actions achieve certain targets ahead of schedule and fulfill their associated goals, the organization can focus more attention on other areas where they are falling behind. By monitoring progress of the portfolio of indicators (sometimes called a dashboard), the organization can change its mix of innovative actions within the innovation process and thus better manage the drive toward achieving its performance targets, goals, and vision for the future.

Balanced Scorecard

The balanced scorecard is regarded by many analysts and practitioners as one of the most effective management techniques in recent decades. The balanced scorecard was developed by Robert Kaplan and David Norton (1996) as an approach to strategic management and associated performance measurement and development initiatives. The model advocates a top-down approach to performance measurement in which a framework structure is used to translate organizational goals into stated objectives and measures. In their research into the way companies measured performance, Kaplan and Norton studied Analog Device's approach, called the corporate scorecard, which involved measuring customer delivery times, product quality, life cycles, and effectiveness in innovation. They revised this approach and introduced new measures such as shareholder value, new compensation plans, and productivity measurements. They integrated them into the measurement system we now know as the balanced scorecard. The balanced scorecard divides strategic objectives, performance measures, and any associated development initiatives into four perspectives (Figure 6.2): financial perspective, customer perspective, internal processes perspective, and learning and growth perspective.

FINANCIAL PERSPECTIVE

Financial measurement remains a key variable in performance measurement but is balanced with nonfinancial measures in the scorecard

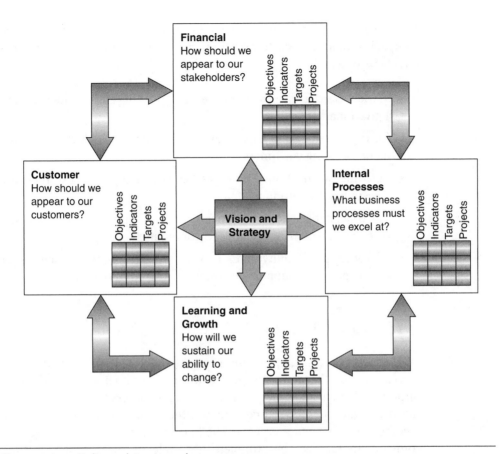

Figure 6.2 Balanced Scorecard

SOURCE: Adapted from Kaplan and Norton (1996).

technique. The balanced scorecard measures both tangible and intangible assets to give an accurate account of the company's physical and nonphysical value. The balanced scorecard encourages organizations to align financial objectives with corporate strategy by selecting objectives and measures that focus on all the scorecard perspectives.

CUSTOMER PERSPECTIVE

Kaplan and Norton found that changing the company's perception of how they value their customers was becoming an important factor in planning and managing performance. Companies needed to become customer focused by introducing strategies to change their objectives. Implementing the customer perspective of the balanced scorecard focuses the organization's mission and strategies into clear objectives. These objectives should include meeting customer demands through quality and growing the customer base and market share. The balanced scorecard encourages the company to focus on five important measures. The company should then customize these five measures to the organization's target customers:

- Market share: This measures customer spending and products sold in a given market.
- Customer retention: This measures the rate at which the company retains existing customers.
- Customer acquisition: This measures how many new customers are investing in the products.
- Customer satisfaction: This measures how happy the customers are with existing offerings relative to competitors.
- Customer profitability: This measures the net profit per customer after expenses to support customers are taken into account.

INTERNAL PROCESSES PERSPECTIVE

The internal business processes perspective focuses on which processes are critical for the company to operate and whether they need to be improved. Measuring these processes allows managers to record how well the organization is operating and whether products and services are meeting customer requirements. The balanced scorecard encourages the company to analyze the complete life cycle of the product, putting emphasis on researching future markets and customer needs, developing high-quality products, and providing ongoing support and services to customers after the sale of the product. Traditional methods of measuring internal processes involved measuring the performance of the process through output, but Kaplan and Norton incorporated the innovation process as a vital component of the internal business process because it allows the organization to invest in research, design, and new developments to compete in emerging markets.

LEARNING AND GROWTH PERSPECTIVE

The learning and growth perspective is focused primarily on employees within the organization. It deals with employee training, which encourages learning, improvement, and innovation to meet the goals and objectives of the organization. The learning and growth perspective provides the platform to successfully carry out the other perspectives in the balanced scorecard. The balanced scorecard stresses the need to invest in the future. In this perspective, investments in people training, systems, and procedures are key to excelling in the future.

Implementing the Balanced Scorecard

A scorecard must be carefully designed so that it leads to an ongoing series of management decisions, actions, and reviews. Kaplan and Norton (1996) outline five principles to achieve breakthrough performance in the organization when using the balanced scorecard.

TRANSLATE STRATEGY INTO OPERATIONAL TERMS

Kaplan and Norton use the term *strategy map* to highlight this process. Each organizational subunit adopts the strategy map that uses the same four perspectives to create their individual operational strategic objectives. These maps help the organizations to identify and select important strategic objectives, associated measures, and then initiatives or projects for implementing strategy.

ALIGN THE ORGANIZATION TO THE OVERALL STRATEGY

Each scorecard in all of the business units and departments can be linked and integrated. The overall strategy should be communicated throughout the organization.

MAKE STRATEGY EVERYONE'S DAY JOB

Employees need to be educated and trained so that they can understand the organization's strategy and contribute to the success of the company through their work. Personal scorecards can be introduced to help the employees adapt to their personal objectives, and reward systems can be put in place to motivate employees.

MAKE STRATEGY A CONTINUOUS PROCESS

Strategy must be linked to the budgeting processes through strategic budgets and operational budgets. Management meetings must be held to review strategy continuously. Ideas generated should be used to constantly fine-tune the company's strategies.

MOBILIZE CHANGE THROUGH LEADERSHIP

Leaders need to lead by example. They must own and be involved in the strategy process so the desired innovation can be driven and accomplished through teamwork. The top-down approach advocated by the framework requires organizations to develop scorecards to reflect organizational demands at different levels (e.g., executive level, business level, department level, project level, individual level). At each stage, the four perspectives of the scorecard are examined and defined in terms of key objectives, measures, targets, and initiatives generated. The successful implementation of the scorecard approach should translate an organization's mission or vision and objectives into a comprehensive set of performance indicators (Kaplan & Norton, 1996). The aim of the scorecard is not to be a control and compliance system but rather to be a system for communicating, informing, and learning.

EXAMPLE: Reflective Display Corp. is an organization that designs and manufactures reflective display films. The design department is engaged in a number of design projects. Its innovation team consists of all department personnel plus people from marketing and manufacturing. The team meets monthly to discuss the execution of its innovation plan, particularly the performance of its indicators. Table 6.1 shows the status of its current performance indicators.

Summary

Performance indicators are a measurable goal for an organization. They can provide a tangible incentive for employees to generate ideas, evaluate progress, and take remedial action as needed. Performance indicators foster improvement and innovation and also present historical data. Indicators can either lead or lag innovation expectations. Indicators can be linked hierarchically to other indicators in the organization so that the set of indicators relevant to any one group is kept to a manageable number. Performance indicators typically are illustrated using performance charts and tables and provide the organization with a feedback loop that allows it to adapt its mix of innovative actions to its emerging results. The balanced scorecard integrates indicators with strategic objectives and projects in organizations. It is unique in that it proposes four strategic thrusts or perspectives: finance, customer, processes, and learning.

Activities

This activity requires you to create a list of performance indicators for your organization for the calendar year. Select these indicators as measures of the strategic objectives that you have defined for your organization. Specify the origin value of the indicator for the start month (e.g., January)

Table 6.1 Performance Indicators at Reflective Display Corp.

Group	Objectives	Measure	Target	Current	Status
Capacity	Effectively use facility	Increase capacity by 10%	220	200	Amber
	Facility customized	Increase capacity by 10%	220	200	Amber
	Increase flexibility of production to allow for customized products	Implement Kanban ordering system	100	0	Red
Compliance	Maintain compliance with ISO regulations	Improve quality of incoming raw materials	3	1	Green
	Improve safety in the workplace	Reduce number of accidents in the workplace	2	8	Green
	Maintain compliance with legislation	Improve quality of incoming raw materials	3	1	Green
Employee Development	Encourage innovative ideas from employees	Employee motivation	39	30	Red
	Reduce absenteeism	Employee motivation	39	30	Red
Environmental Awareness	Increase proportion of recycling	Reduce waste produced	1	4	Green
	Reduce packaging	Reduce waste produced	1	4	Green
	Reduce waste produced	Reduce waste produced	1	4	Green
Improve Productivity	Improve stock management	Implement Kanban ordering system	100	0	Red
	Reduce lead times	Reduce lead times by 10 days	5	15	Red
	Increase operator flexibility	Reduce lead times by 10 days	5	15	Red
Quality	Reduce defective production on shop floor	Reduce % defects	4	10	Red
	Guarantee quality in incoming raw materials	Improve quality of incoming raw materials	3	1	Green
	Improve work instructions	Reduce % defects	4	10	Red

and the target value for the end month (e.g., December). Each strategic objective can have one or more indicators, and similarly any specific performance indicator can be linked to the achievement of one or more objectives. Copy Table 6.2 into a spreadsheet and complete the fields. *Unit* refers to the unit of measurement (e.g., hours/unit). When defining the specific indicator, record the associated origin and target values. Try to place an active verb in the title of the indicator (e.g., *Increase, Decrease, Maintain*). Next, define who in the organization is responsible for the achievement of this particular indicator. Place a green, amber, or red symbol in the status field to indicate progress toward the performance target. The second part of this activity requires you to complete a simple chart for one of your performance indicators. A sample chart is illustrated in Figure 6.3.

STRETCH: Other elements of this activity may include discovering emergent indicators for organizations in your domain and creating a separate list titled "Emergent Indicators." These emergent indicators can be found in various trade magazines, online discussion forums, benchmarking, and so on and reflect general changes occurring across many organizations. These emergent indicators may or may not result in a revision of the indicators previously defined. Another stretch to this activity is to bring your strategic objectives and indicators together into a modified balanced scorecard table. Use Table 6.1 to begin creating your balanced scorecard. Replace the words in the group column with the four thrusts defined by Kaplan and Norton.

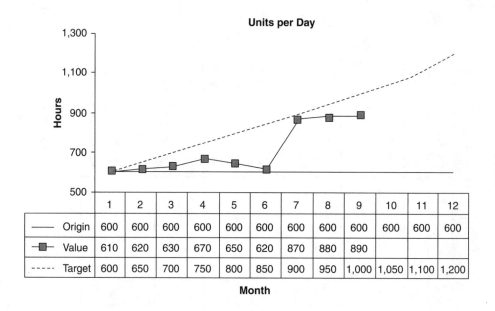

		1	2	3	4	5	6	7	8	9	10	11	12
——	Origin	600	600	600	600	600	600	600	600	600	600	600	600
—■—	Value	610	620	630	670	650	620	870	880	890			
- - - -	Target	600	650	700	750	800	850	900	950	1,000	1,050	1,100	1,200

Figure 6.3   Sample Chart

Table 6.2 Create Indicators

Indicators					
Title	Unit	Current	Target	Responsible	Status

Title: Title of the indicator (e.g., "Reduce defects per unit")

Unit: Unit of measurement (e.g., hours/day)

Current: Current value of the indicator (e.g., 230)

Target: Target value of the indicator at the end of the planning period (e.g., 260)

Responsible: Person responsible for reporting the status of the indicator

Status: Status of the indicator (e.g., not started, red, amber, green, completed)

REFLECTIONS

- What is the difference between a financial and a nonfinancial indicator?
- Why are performance indicators important for motivating employees?
- Explain the balanced scorecard technique.
- List and define three current or emergent indicators used by many organizations.
- Detail a simple form for capturing critical data for an indicator.
- Indicate the key data points in a performance chart.

Part III

Managing Innovation Actions

Actions are the activities an organization carries out to make innovation happen to products, processes, and services. Actions usually are carried out in response to a stimulus such as a problem, a goal, or simply a creative thought. There are a number of different actions, including solving problems, generating ideas, managing projects, and balancing project portfolios. Each of these actions influences the overall success of the innovation process. Problems and ideas can be either reactive or proactive. They can react to an existing problem, or they can be proactive in responding to a potential future problem, an opportunity, or a future goal. Ideas that don't require significant resources can be implemented as "quick wins," whereas those that require more substantial resources can become projects. Projects are unique nonpermanent activities that require resources: time, money, and labor. A typical organization will have many projects active at once and at various stages of implementation. These projects form a portfolio, where they can be ranked and managed to maximize their contribution to achieving goals under the normal constraints of resource availability. All projects will have a relationship to the goals of the organization in order to move it toward its vision of the future. In this part of the book, Chapter 7 looks at how ideas can be created and processed through an action pathway. Some ideas will grow to become projects, and Chapter 8 looks at managing individual projects. Projects related to new products and particularly breakthrough products are a special case of project management, and this is discussed in Chapter 9. Issues such as protection of intellectual property and exploitation are also explored. Chapter 10 looks at a group of projects, or a portfolio, and how to optimize the balance of projects to achieve organizational goals (see Figure III).

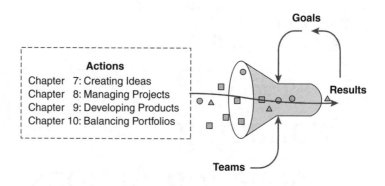

Figure III   Managing Innovation Actions

LEARNING TARGETS

When you have completed this part you will be able to

- Describe the creativity process
- Understand the various sources of ideas used in innovation
- Apply a number of idea generation tools
- Identify core aspects of project management
- Explain the product development process
- Understand how to capture critical data for innovation projects
- Explain the process of project portfolio management

Creating Ideas 7

Innovation is about setting goals and then executing various actions such as ideas for creating change that will achieve those goals. Innovation is also about solving problems. From time to time problems arise in products, processes, and services that need to be corrected. Problems and ideas are identified by employees, customers, and other stakeholders, who can use their experience by playing a key role in the development and implementation of new ideas. These creative concepts provide the starting point for the process of developing product, process, and service innovations. Although the development of each innovation is unique, all pass through a series of phases, or a development pathway, controlled by decision points or stage gates. This action pathway progresses concepts from ideas into more tangible projects that are developed into eventual innovations beneficial to the organization.

LEARNING TARGETS

When you have completed this chapter you will be able to

- Explain creativity
- Explore a number of tools that can be used for generating ideas
- Understand where many ideas come from
- Explain activity modeling
- Understand some problem-solving techniques
- Detail a simple form for capturing critical data for an idea
- Explain the importance of lead users in generating new ideas

Action Pathway

Prospective innovations begin as problems or ideas and flow along the action pathway. As they flow along, they are developed through stages such as opportunity recognition, development, realization, and learning (Figure 7.1). Each stage of the pathway requires a decision to move to the next stage, with each subsequent stage requiring more resources (time and money) to execute the idea. Further development provides the organization with more knowledge about how the opportunity can be realized. With richer information, the organization can make better decisions on whether to progress or abandon the idea. Organizations face one of the fundamental dilemmas of innovation management: the wish to minimize the amount of money expended on poor ideas while also realizing the need to invest in ideas in order to determine true potential. As an idea flows through the various phases of the process, the cost can increase significantly. Many managers promote the mantra of "fail fast and early" to encourage themselves to abandon unattractive ideas as early as possible and allow scarce resources be expended on more worthwhile ideas. Although the pathway is often represented as a linear process, this is a simplification of reality. Many of the stages overlap with each other, and various actions can loop back to earlier stages. For the remainder of this chapter, we will focus on the initial stage of the action pathway and examine ways in which the organization can create ideas that will lead to eventual innovations or full implementation that adds value to customers.

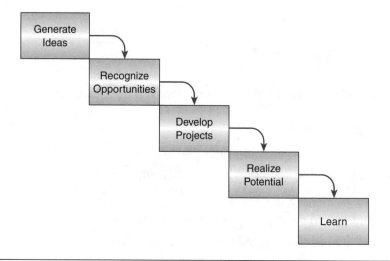

Figure 7.1   Action Pathway

Problem Solving

Current or future problems that may exist in the product, process, or service may form one potential source of ideas. The problem-solving process can be said to involve a number of steps, from identification of the problem to solution development, implementation, and follow-up. The steps are as follows:

1. Identify and select the product, process, or service to be improved.

2. Ensure that the right resources are available and objectives set.

3. Identify problems, prioritize, and select specific problems to analyze.

4. Identify and verify the key causes of a problem.

5. Identify and verify alternative solutions to key causes.

6. Select and check potential solutions.

7. Implement the identified solution.

8. Review the success of the solution implementation.

9. Follow up on failures and identify improvements.

Everyone in the organization can participate in problem solving, from senior management down to operatives on the shop floor and front-of-office staff who deliver services to customers. The best way to identify and solve a problem is to empower and enable each person to examine his or her own job function and solve problems or deficiencies systematically. People need to have the skills and motivation to identify and solve problems. A number of problem-solving tools and techniques can be used by individuals and groups for this process. These include cause–effect, brainstorming, and ranking. Problems that may potentially exist in the future can also be tackled using techniques such as failure mode effects analysis (FMEA). Problems typically will lead to incremental changes to existing products, processes, and services but under certain conditions can stimulate more radical innovation.

Creativity

Creativity is essential if a concept is to be novel and useful and offer real competitive advantage to the organization. Creativity is often identified by the presence of inspiration, cognitive leads, and intuitive insights that result in the addition of originality to the innovation. Two conflicting views regarding the origins of creativity are the genius view and the behaviorist view. The genius view proposes that creativity is the domain of geniuses such as Leonardo da Vinci, a direct gift from the gods. The behaviorist view proposes that creativity is "the universal heritage of every human being" (Maslow, 1962, p. 167), and like any skill, it can be developed. Organizations that focus only on trapping genius creativity are ignoring the significant imaginative thought that is present in all people. The enlightened perspective of creativity in the context of organizational innovation is that it is the domain of all people rather than being possessed by the lone genius. A creative output is the result of the natural thought processes of ordinary people. Therefore, all people have the potential to contribute to creative concepts if provided with the correct environment and training. In the organizational context, not all creativity is desirable; only ideas that are appropriate to the organizational context and are "useful and actionable" are desirable because it is these creative concepts that can improve organizational products, processes, and services (Amabile, 1998, p. 77). The creative output must be managed through appropriate training and communication to ensure that what is produced is desirable in light of the goals of the organization. The creative capability of an organization can be viewed as the synthesis of the specific person, the specific task, and the specific organizational culture. Individual creativity has been discussed at length by various authors in an attempt to identify the typical traits that a creative person needs. Although traits such as openness to new ideas, curiosity, persistence, autonomy, and self-reliance contribute to creative capability, no conclusive profile has been developed. However, there are some very useful insights.

One useful way of framing the creative capability of an individual is to view it as the sum of three components (Amabile, 1998):

- Expertise
- Creative thinking
- Motivation

A person's level of creativity is a function of these three components. Expertise is the technical and intellectual knowledge people possess and the manner in which the organization manages this collective knowledge. Typically this is a learning process for the individual over a long time frame. Creative thinking involves the person's technical and procedural skills at problem solving and idea generation. Again, these skills can be learned but usually over a much shorter time frame. Motivation results from the intrinsic and extrinsic factors influencing a person to be creative. The three components are interrelated, and development in one area can exert a significant effect on the other two. On the other hand, development in one area does not rule out the need for development in the other two. The work environment can contribute significantly to increasing expertise, creative thinking, and motivation in an organization and ultimately affect the creative output.

Another perspective on creativity looks at some of the ways in which creative ideas come into being. Goffin and Mitchell (2005) propose three different types of creativity through which ideas come about: normative, exploratory, and serendipitous.

NORMATIVE

Normative creativity occurs when original thinking is applied to solve an existing problem. An example of this type of creativity occurs when people in an organization identify more efficient processes in response to customer demands and requirements.

EXPLORATORY

Exploratory creativity involves ideas that fundamentally challenge the existing norms and routines in the organization. This type of creativity is often more radical in its impact, and it can encounter significant resistance as it threatens existing practices and power structures. It usually involves significant time and investment on the part of the organization to build up the competence necessary to allow "structured breakthroughs." An example of this type of creativity is when a pharmaceutical company develops a new drug for the market.

SERENDIPITOUS

Serendipity is creativity that occurs through an act of unintentional good fortune. An example of this type of creativity is the discovery of penicillin, when contamination of a petri dish led to Alexander Fleming's remarkable scientific breakthrough.

Each of these types of creativity contributes to innovation and clearly offers benefits to the organization. Creativity is relevant not only at the beginning of the innovation process but also throughout development, right up to the realization of the innovation. All creative concepts face challenges that must be overcome if they are to become successful innovations, so creativity is as essential in these later stages as it is in the ideation phase. These challenges normally relate to technical difficulties but can also include issues such as overcoming resource constraints or reacting to emergent shifts in the external environment.

EXAMPLE: In the 1920s 3M sold sandpaper products. A young designer observed problems being experienced by painters in a car shop when they tried to put a colored stripe on a body panel. They used tape to blank out the stripe on the panel when putting on the first color. This was then removed, and a second tape was placed over the edges of the first color before they painted the stripe. When the second tape was removed, some of the first color came away with it. It was necessary to either restart the process, which took too long, or touch up the paint work, which affected the surface finish of the product. The young designer considered the problem and was determined to create a less sticky tape. His boss told him it was outside their scope, but the designer ignored this and worked weekends to create a solution. His boss decided to be patient and gave him some free time from his normal duties. The designer eventually created masking tape, which became one of 3M's biggest-selling products. His boss learned from the experience and decided to implement an innovation culture that mirrored many of the traits mentioned in this example: taking risks, allowing time to experiment, observing customers, challenging authority, and so on. 3M went on to become one of the largest corporations in the world, with many different types of products.

Enhancing Creativity

Creativity is often seen as an individual act in which one person creates an idea, which is then developed and exploited by the organization. However, most innovations are the result of team efforts. Teams can create more innovation because they bring together different competencies, insights, and perspectives. Team composition means a diversity of thinking styles and skills that can result in more imaginative and robust ideas being generated. This diversity has a number of advantages:

- Diversity creates creative friction between individuals that can spark new ideas.
- Diversity is a safeguard against groupthink, in which a group of people allow their thinking to converge over time.
- Diversity creates an environment in which different perspectives are developed and good ideas can be identified, tested, and supported.

In order for teams to be both creative and effective, they must be able to balance a number of paradoxical characteristics (Luecke, 2003). These conflicting characteristics include the need for naïveté versus experience, autonomy versus discipline, fun versus professionalism, and improvisation versus planning.

NAIVETY VERSUS EXPERIENCE

In order to be creative, teams benefit from having members of diverse backgrounds and experience who can analyze problems from different perspectives. Having team members experienced in the problem area provides access to a rich reservoir of expert knowledge. Similarly, having members who are naive regarding the established norms will challenge these perceptions and avoid groupthink. By achieving an appropriate balance between naivety and experience, the team can produce a larger and broader range of ideas.

AUTONOMY VERSUS DISCIPLINE

Another paradox that must be managed in order to maximize the creative output of the team is the balance between autonomy and discipline. Teams are established to deliver specific objectives that relate to organizational goals, and therefore there must be a structure for ensuring that these deliverables are achieved within the agreed time frame. However, teams cannot be micromanaged by the organization and need a high level of autonomy in order to achieve their task. Given the complexity of the challenge, teams need the freedom to be agile and imaginative in their delivery of the desired result for the organization but disciplined enough to achieve results within anticipated time frames.

FUN VERSUS PROFESSIONALISM

As teams strive to achieve their objectives, they must balance the need for successful professionalism with the need for fun and experimentation in order to develop creative solutions. The balance that the team achieves between these two contradictory characteristics will influence the team culture and the level of motivation exhibited by team members.

IMPROVISATION VERSUS PLANNING

The final paradox that must be addressed is the balance between planning and improvisation. As the team strives to achieve their objectives, they will need to plan the use of the available resources and skills in order to operate efficiently and deliver in a timely manner. Although effective planning is crucial to success, the team must also be able to adapt to changing circumstances.

Encouraging Creativity

Creativity begins by identifying a problem or opportunity. Ideally, individuals or teams with the right expertise, motivation, and creative thinking skills work on a problem or opportunity; they generate ideas, test possibilities, and ultimately implement actions such as quick wins, corrective actions, or projects. Most ideas will be scrapped, recycled, merged with other ideas, or postponed as they progress along the action pathway. Perhaps only one idea in one hundred will progress to become a solution, and therefore organizations need to ensure that they have a plentiful supply of ideas feeding into the mouth of their innovation process (i.e., the innovation funnel). There are a number of ways of encouraging problem solving and idea creation in any organization. Some of the more popular ways are as follows:

- Providing good strategic direction
- Benchmarking and access to external stimuli
- Providing a diverse information service
- Employing staff with diverse interests
- Having a supportive management style
- Creating a climate for innovation
- Allowing failures to be tolerated
- Allowing people to pursue their own ideas
- Rewarding success
- Interacting with the customer
- Collaborating with other organizations and lead users
- Encouraging cross-pollination of ideas
- Providing idea suggestion programs

Creativity is important not only because of its contribution to organizational goals. Certain creative outputs are valuable because they clash with the organization's goals and challenge conventional thinking. This type of creativity can cause an organization to change the goals it is pursuing in order to shift its trajectory or develop a concept that has the potential to revolutionize the industry by creating a disruptive shift. This "creative destruction" (Schumpeter, 1942) can offer an organization the opportunity to wipe away the power structure of the existing industrial environment and rewrite the rules of competition.

Sources of Ideas

Moving from the macro to the more specific, the sources of innovation can be viewed as diverse. Innovation can originate from employees' experiences and interests and from the organization's existing knowledge and process capabilities (Smith, 2006). Although these can be viewed as internal to the organization, external sources also offer opportunity for innovation. Innovations can originate from customer routines and practices, from spillover of knowledge and technology from other industrial sectors, from lead users, from universities, and from supplier organizations. Sources of innovation can be divided into six areas (Luecke, 2003): new knowledge, customer ideas, lead users, empathetic design, invention factories, and open innovation (Figure 7.2).

NEW KNOWLEDGE

These are ideas from employees, suppliers, distributors, and individuals in the extended organization. Ideas typically are generated from new knowledge and insights gained from books, magazines, competitive benchmarking, collaborative benchmarking, research, practice, and experience. Many sources of new knowledge can be used to generate new ideas. This is explored in more detail in the next section.

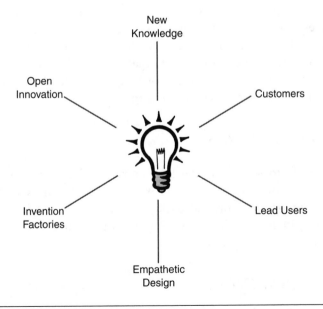

Figure 7.2   Sources of Ideas

CUSTOMER IDEAS

Customers are a main source of new ideas for products, processes, and services. Customers are useful for identifying problems and weaknesses. Market research of large customer bases can also identify future trends in customer buying behavior. One weakness of using customers solely is that they often defend the products they purchase and therefore usually have poor ideas about future products, processes, or services that may make their current purchase obsolete or that question their judgment in making the purchase in the first place.

LEAD USERS

Some customers face greater needs than others and are interested in engaging in the innovation process in order to push the barrier of usage for particular products, processes, and services. These are called lead users and are valuable sources of innovation because they encounter needs far in advance of the majority of the market. Thus, lead users provide foresight into potential mass market needs and are important for collaborative benchmarking, co-design, testing, and validation of new ideas. Many organizations value the opinions of these lead users to the extent that they go to great trouble and expense to solicit their views and observations.

EMPATHETIC DESIGN

This design process involves observing users of the products, processes, and services. Users often are observed by camera over a prolonged period. Their usage pattern often shows pleasure, frustration, and so on— reactions that can guide designers in regard to people's preferences for a particular product or service. This knowledge can be used to guide future designs of products, processes, and services.

INVENTION FACTORIES

Invention factories are special laboratories within organizations and those shared by organizations and universities. Laboratories such as Bell Labs hire experts from diverse backgrounds to work on ideas for the future. These ideas are principally scientific, with solutions that are often possible only in the long term.

OPEN INNOVATION

This approach recognizes the innovative capability that exists outside the organization. By developing systems that allow suppliers, competitors, customers, and even members of the public to engage with the organization's innovation process, a much broader scanning of the external environment for opportunities can be achieved. Similarly, the organization is able to access the skills and capabilities of these collaborators in order to support the development of specific innovations. The concept of open innovation views the organization as an enabler and facilitator among a number of independent organizations. One particular example of this approach to innovation is that of the Linux operating system: The code is open source, and programmers from around the globe add new enhancements to the communal product.

New Knowledge Ideas

Many ideas come from within the organization in the form of new knowledge. Other ideas are reported in books and magazines as being the latest trends in new technology and techniques. These ideas often are common concepts that can be of interest to a wide variety of organizations. Organizational ideas can evolve around application of technology, engagement of employees, improved customer interaction, benign environmental impact, and cost reduction.

APPLICATION OF TECHNOLOGY

Many organizations use technology as an enabler to enhance their ability to share knowledge better, improve process efficiency, and enhance product offerings. Ideas such as the integration of databases, the commissioning of more productive manufacturing equipment, or the exploitation of novel technology to enhance product performance can contribute to the organization's innovative effort.

ENGAGEMENT OF EMPLOYEES

Organizations strive for higher engagement of employees in both the day-to-day operations of the organization and the innovation process itself. Ideas such as novel reward systems, enhanced training systems, and increased empowerment can boost employee loyalty and commitment and can result in a stream of novel ideas for innovation.

IMPROVED CUSTOMER INTERACTION

Many organizations seek ideas that break down boundaries between themselves and the customer in order to enhance customer satisfaction.

Ideas about integrating quality tools such as quality function deployment, conjoint analysis, and rapid prototyping into the development process can enhance both the efficiency of the process and the quality of the products that are delivered to the market. Similarly, focusing on the internal customer, the application of techniques such as concurrent engineering and co-design and the altering of organizational structures to adopt a more team-based approach are popular emerging concepts.

BENIGN ENVIRONMENTAL IMPACT

With increased emphasis on environmental issues, many organizations are embracing ideas that will improve the organization's environmental credentials. These ideas usually focus on reducing the carbon emissions caused by production; this is done by increased recycling of the physical product through enhanced design and use of alternative raw materials or reprocessing of waste byproducts in an environmentally friendly way.

COST REDUCTION

Many concepts focus on how the organization can reduce its overall costs in order to increase its margins and sustainability. Ideas such as the use of alternative raw materials for products, alternative production and distribution models, or even alternative business models can offer the organization the opportunity to reduce its costs. This is perhaps best illustrated through the lean initiatives adopted by most manufacturing facilities and increasingly by service organizations.

EXAMPLE: Theme Park Design Group is an entertainment company that markets, designs, and operates medium-sized theme parks. The design group department regularly receives demands for new features from marketing and other departments. It keeps these ideas in a simple information system. The group has a number of ideas in its portfolio (Figure 7.3) and has ranked them according to their level of impact and risk. Some of these ideas have been translated into projects.

Activities

This activity requires you to create a list of about 10 fictitious ideas for your organization. These creative concepts can come from your environmental analysis, goal definition, benchmarking, or creative abilities. Some of these ideas may be translated later into particular projects.

Idea ⬍	Responsible ⬍	% Complete ⬍	Status ⬍	Decision ⬍
Career Development Plan	Calvin	50	☼	Draft
Design a computer game section within a park	Costello	30	◎	Draft
Allow staffs to work flexibly	Calvin	15	☼	Draft
Improve the department management system	Costello	20	◎	New Project
Purchase new computers from Apple company	Costello	99	✓	Merge
Recruit experienced designers from other companies	Calvin		◉	Draft
Simplify design specification documents	Calvin	0	✓	Abort
Training in safety design	Calvin	80	◎	Merge
Extend a Korean history section in Seoul Theme Park	Giulia	99	✓	Merge
Simplify the operation steps of a certain rider	John	0	☼	Merge

◉ Requires urgent attention   ◎ Progressing satisfactorily

☼ Requires Attention   ✓ Completed

Figure 7.3   Idea Portfolio for Design Department

Group the ideas according to whether they began as a problem or from new knowledge, stakeholder requirements, or other stimuli. Once you have recorded these ideas, assess the suitability of the concept by scoring their impact and risk on a scale of 1 to 5. Based on this analysis of the idea, rank the priority of the ideas using a 1 to 5 scale. Assign responsibility to a suitable employee and set a due date for the idea to be completely investigated and returned to the team for further review. Keep the number of words to a minimum. Copy Table 7.1 into a spreadsheet and complete the fields.

Once the idea investigation is completed, the new information can be assessed to decide whether the idea should progress further along the action pathway. Possible decisions can include "Approval as project," "Reject following investigation," "Place on hold," or "Return for further investigation."

Now repeat this activity to create a list of approximately five fictitious problems for your organization. Some of these problems may be later translated into particular ideas or projects. Group the problems into, say, "Reactive" and "Proactive." Copy Table 7.2 into a spreadsheet and complete the necessary fields.

STRETCH: As you continue through the other exercises in this book, revisit both of these activities periodically and add new ideas and problems to the two tables you have created. The innovation process is ongoing and should be continuously updated with new creative concepts.

Table 7.1 Create Ideas

Ideas									
Group	Title	Impact	Risk	Priority	Due	Responsible	Status	Notes	

Group: Title of the idea group (e.g., "Problems" or "Suggestions")

Title: Title of the idea

Impact: Impact of the idea on goal attainment from 1 to 5

Risk: Level of risk associated with the idea in achieving its impact from 1 to 5

Priority: Priority of the problem from 1 to 5

Due: Due date for completing investigation of the idea

Responsible: Person responsible for investigating the idea

Status: Status of the idea (e.g., "not started," "in progress," "waiting," "completed")

Notes: Additional information such as associated project name or reference

Table 7.2   Create Problems

Problems									
Group	Title	Impact	Risk	Priority	Due	Responsible	Status	Notes	

Group: Title of the problem group (e.g., "Reactive" or "Proactive")

Title: Title of the problem

Impact: Impact of the problem of goal attainment from 1 to 5

Risk: Level of risk associated with the problem occuring again from 1 to 5

Priority: Priority of the problem from 1 to 5

Due: Due date for completing investigation of the problem

Responsible: Individual responsible for investigating the problem

Status: Status of the problem (e.g., "not started," "in progress," "waiting," "completed")

Notes: Additional information such as associated project name or reference

145

Investigating Ideas

The first stage in the action pathway is idea generation. The idea generation stage can be said to consist of a number of substages: preparation, incubation, illumination, verification, and elaboration (Kao, 1989). Irrespective of whether creativity is being used to solve a problem or generate ideas, the process is essentially the same. Each substage can be supported by the use of a number of tools and techniques that facilitate problem definition, idea generation, idea ranking, idea selection, and eventual idea implementation (Flynn, O'Sullivan, & Dooley, 2003). Ideation tools include :

> *activity networks, affinity diagrams, bar charts, brainstorming, cause–effect diagrams, checklists, control charts, decision trees, design of experiments, fault tree analyses, failure mode effects analyses, flowcharts, flow process charts, force field diagrams, Gantt charts, histograms, line charts, matrix diagrams, matrix data analysis charts, nominal group techniques, Pareto diagrams, prioritization matrices, process capability diagrams, process decision program charts, relationship diagrams, scatter diagrams, string diagrams, surveys, tables, tree diagrams, value analyses, and voting.*

Each of these tools provides a particular technique for organizing data. Often the structure helps to clarify a problem or idea and permits the data to be interpreted more easily. The structure satisfies the key need for reliability in the ideation process, which will facilitate reproducing the process. Structure also provides a framework for understanding the main goals of ideation. The degree of confidence that can be placed in a tool often is reflected in the level of structure it achieves. For example, a tool such as brainstorming minimizes structure and is primarily effective in breaking down barriers. Choosing the right set of tools requires a good understanding of the idea or problem and a good understanding of the use and application of the tool. Time and effort must be invested in defining the root objectives of a problem. Even when the decision seems clear-cut, it is worth pausing to make sure that the right problem or objective is being addressed. This will ensure that the right tool is picked for the job. It is not possible to discuss each of these tools individually. However, in the next section we highlight some of the more widely used tools that have proved useful in engaging all levels of the organization in the ideation process. You are encouraged to undertake a more detailed review of any of the tools mentioned in the list above using the Internet.

Selected Ideation Tools

A number of tools are popular for defining problems, discovering solutions, and generating ideas. Their popularity stems from the fact that they are easily understood, are easy to apply, and engage all levels of skilled individuals in the organization. This section reviews the following popular tools: cause–effect, brainstorming, ranking, failure mode effects analysis, and mind mapping.

CAUSE–EFFECT

This graphic technique identifies the effect of a problem and then examines all possible causes of the effect. Causes can be grouped into key categories. One popular grouping technique is to use the 4 Ms: *man, machine, method,* and *material.* Under this technique, the effect is first noted (e.g., "late product releasse"). The possible causes of the effect are then examined under the four main headings. Are there possible causes for the failure in the human (man) or perhaps the machine used to produce the product (machine), or the method used in producing the product (method), or the materials used in the product (materials)? At the end of this exercise, a list of possible causes of the effect under each of the headings is produced and mapped. This output then becomes the basis of brainstorming exercises to develop potential solutions to a number of the most likely causes (see Figure 7.4).

BRAINSTORMING

Brainstorming is a technique that relies heavily on group creativity. It is particularly effective in looking in the broad direction of a problem and developing solutions to problems that cannot be logically deduced. Brainstorming encourages the use of divergent thinking. Two basic rules apply to brainstorming: Judgment is suspended during the creation of

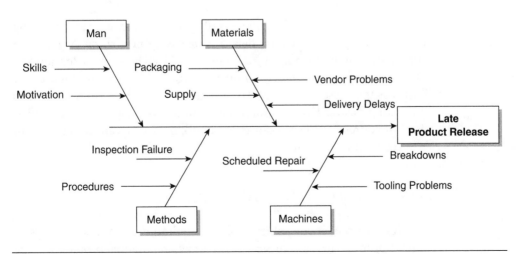

Figure 7.4   Cause–Effect Diagram

ideas, and all ideas put forward are considered. The technique usually is undertaken by a group of people of diverse backgrounds who discuss possible solutions to a particular problem (or cause of a problem). The group generates as many ideas as possible around a central theme and documents their output. The group should avoid assessing the feasibility or practicality of any particular idea during this exercise.

RANKING

Once a list of ideas has been generated in the brainstorming exercise, the next possible step is to rank them in order of preference. Ideas can be ranked in numerous ways. One popular approach is to rank them according to risk and impact at solving the problem or effect. Each person can be asked to place a number from 1 to 5 for both parameters on each idea. Alternatively, an open discussion between groups can attempt to reach a consensus on the ranking. A score for each idea is calculated by multiplying the average risk and impact scores. Once the ideas have been ranked, perhaps the top three to five can be discussed further in detail and, if deemed appropriate, forwarded to more senior management for implementation approval. Evaluating ideas can also be effective in judging merit with respect to impact on the goals of the organization. Other factors that might be considered include the technical and business competencies available in the organization. There is little point in choosing a good idea if the skills to implement and exploit the idea are not available. Cost–benefit analysis is another direct way of evaluating an idea, although it is often difficult to determine benefit from an idea that has not been tested in the marketplace. Irrespective of the method used to rank the various ideas, the most important aspects are consistency and accuracy of selection.

EXAMPLE: Customers and employees at a computer chip manufacturing company complained of long delays in releasing new chips. A group of employees including managers, designers, and lead customers developed a solution using cause–effect, brainstorming, and ranking. Figure 7.4 illustrates the cause–effect diagram, which indicates the most likely causes of the long delays. This was followed by brainstorming, which identified ideas for addressing the key causes, and ranking, which ranked the ideas in terms of their risk and impact.

FAILURE MODE EFFECTS ANALYSIS

FMEA is a proactive problem-solving method used to identify ways in which a product, process, or service might avoid potential problems in the future. FMEA also aims to determine the effects of potential failures on performance. The process ranks failure effects and causes. This ranking indicates where design effort should be concentrated to reduce future failure occurrence. The technique draws essential experience and information from design, marketing, production, purchasing, and distribution in order to pinpoint the critical nature of potential problems and suggest preventive action. There are three main elements in FMEA: failure effect, failure cause, and failure criticality.

Failure effect: The potential failures are studied to determine their probable effects on the product, process, or service and the effects of various components on each other.

Failure cause: The anticipated conditions of operation are used as the background to study the most probable failure causes.

Failure criticality: The potential causes of failure are examined in order to determine their severity in terms of lowering performance, causing a safety hazard, total loss of function, or other problems. Each cause is given an RPN value: R is the level of risk, P is the probability of the failure occurring, and N is the difficulty of detection. The RPN value is the product of $R \times P \times N$.

The causes of failure with the highest RPNs are developed further in terms of identifying design changes that can lower the chances of failure or prevent it altogether. FMEA has been successfully applied to product, process, and service design.

EXAMPLE: Thermo King recently used FMEA in one of its assembly lines. Each assembly line is divided into cells that have one or two operators and a number of specific assembly tasks. Each task has been given a duration for completion of the task and a clearly defined RPN. Some tasks have higher risks, and the operator can see this on the task list. Other tasks have led to failure in the past, and this is reflected in the frequency of the occurrence of failure or probability value for the task. Finally, each task is also assigned a value for difficulty of detection. When combined, these values highlight to the assembly line operator the relative importance of each task toward future failures and hence the relative caution needed in ensuring successful completion of the task. The times allocated to each task are also factored up if the task has a high RPN. The implementation of FMEA yielded a 35% reduction in assembly-related failures. There were significant cost savings both in the assembly process (through reduced reworking of poor assemblies) and in after-sales warranty costs.

MIND MAPPING

Mind mapping has become a popular tool in recent years for a wide variety of problem-solving and idea generation tasks. Mind mapping, which takes its ideas from the original spider diagram, can be used to create concepts, associations between concepts, and hierarchies of concepts. Variations of mind mapping include tree diagrams and topic maps. Radical tree diagrams, hierarchical tree diagrams, and clustering methods all use the same hierarchical logic. Various software packages are available that support the generation and editing of mind maps. Some software tools can also accommodate attachments such as notes, hyperlinks, and documents.

Mind maps are centered on a core concept, such as publishing this book. Related concepts are then developed around the core concept such as the chapters, the potential publishers, and the potential audience. Each related concept can then be grown further to lower-level concepts. For example, the concept of chapters can be divided into core text, figures, activities, and so on. It may be appropriate to put the same item in more than one place. It may also be necessary to show relationships between items on different branches. The approach at first seems trivial, but mind maps are an aid to mental mapping of concepts. Mind maps rarely have meaning outside the individual or group that created them. They act as a decision support tool for the idea generation process. An example of a mind map is given in Figure 7.5.

Modeling Tools

So far we have focused on tools that can be used to deal with specific problems or ideas. Often a tool is needed to allow individuals and teams to gain a broader understanding of the organizational system as a whole. These tools can include process modeling, data flow modeling, and discrete event simulation modeling. One of the most common tools for modeling and analyzing processes or organizational activities is IDEFo. IDEFo is an activity modeling method for modeling mainly process activities. It is a powerful way for teams to model the behavior of the whole organization before generating ideas for change. Its power often comes more from the

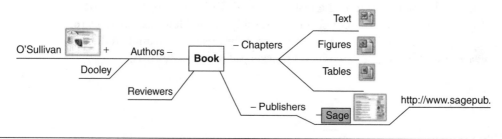

Figure 7.5 Mind Map

learning process during model development than from the model itself when completed. An IDEFo model is constructed using two basic modeling principles: cell modeling graphics and hierarchical decomposition.

CELL MODELING GRAPHICS

IDEFo models the activities and flows in an organization with rectangular boxes and arrows. Boxes represent activities, and arrows represent the input and output flows. Because they are activities, the text label inside the box must use an active verb to describe the activity (e.g., "make goods," "implement program," "operate machine"). The way in which the arrows enter the box is important. Arrows entering from the top are controlling inputs. These inputs control the execution of the activity (e.g., customer orders control the work done in a machine cell). Arrows entering from the left are simple inputs; they do not control the activity but are transformed by it into the outputs (e.g., raw material input is transformed into finished goods), which exit from the right of the activity. Finally, arrows entering from the bottom represent resources needed for the activity (e.g., a machine is needed to produce a part from raw materials). Figure 7.6 illustrates these types of flows between two activities.

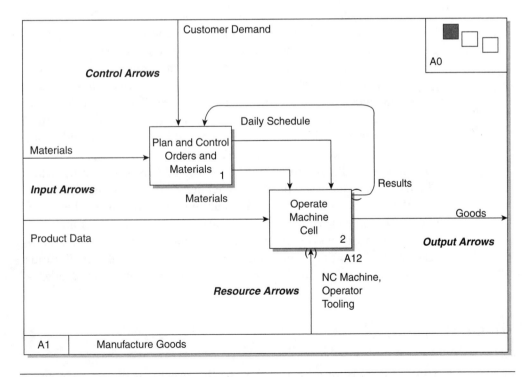

Figure 7.6   Activity Flows

HIERARCHICAL DECOMPOSITION

Another important approach used in IDEFo is the stepwise revealing of detail through hierarchical decomposition. Each activity box in a model represents a number of potential subactivities, and each arrow represents a number of subarrows. New diagrams can be created for each box, and these new diagrams contain even more boxes that detail the activity being modeled. The concept of hierarchical decomposition is illustrated in Figure 7.7.

BUILDING AN IDEFo MODEL

In general, IDEFo modeling adopts a six-stage approach:

1. Select viewpoint and purpose for the model.
2. Create a one-box diagram (A-0) for the activity.
3. Create a multibox diagram (A0) for the activity.
4. Create a parent diagram for the one-box A-0 diagram (A-1).
5. Create subdiagrams for the multibox diagram (A1, A1.1, etc.).
6. Review model and check for purpose and viewpoint.

The development of an IDEFo model depends on the author's purpose and viewpoint and the degree to which he or she adheres to the general guidelines laid down in the IDEFo method. There are many benefits in adhering to these guidelines, but in general some shortcuts can be taken, particularly with respect to the reader–author cycle, in which the creators of the model request feedback from other people regarding the accuracy of the model. IDEFo is a comprehensive system modeling and analysis method. The technique is in widespread use throughout manufacturing and is simple yet rigorous enough to satisfy system analysis requirements. However, it does have limited application to the conceptual levels of system modeling. There are a number of modeling guidelines or rules that, used together, help the IDEFo author to produce accurate representations of the subject matter. They include conciseness, gradual exposition of detail, limitation to no more than six boxes per diagram, diagram interface connectivity, uniqueness of labels and titles, minimum control of function (all activities must have a control arrow), and a clear purpose and viewpoint for the model.

Physical Space

Another important aspect of the creative process is the physical space in which people work. Many innovative organizations recognize the

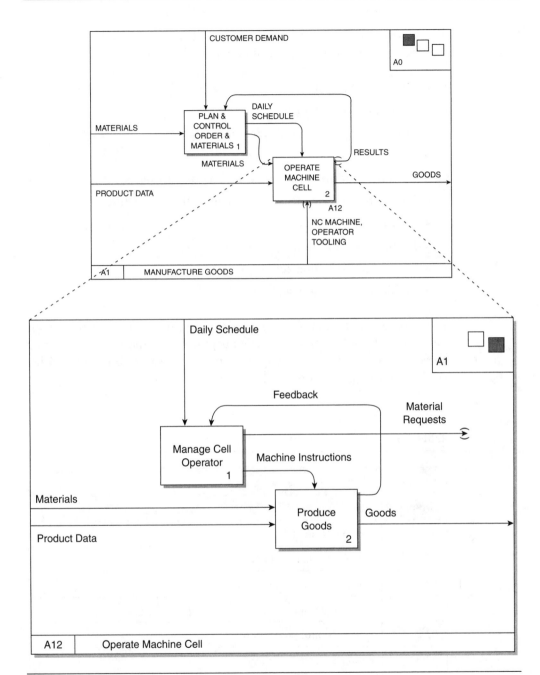

Figure 7.7 Hierarchical Decomposition

importance of investing in physical workspace. A creativity-friendly work-space is one that contains many different types of stimuli and is physically laid out to make it easier to promote idea generation and sharing between individuals. When physical workspace cannot be found, for example, where people are not located in the same building, virtual workspaces can be used to improve creativity. A number of measures can be adopted to improve physical workspace, including the following:

- Casual meeting areas
- Customer contact areas
- Libraries (e.g., with books, reports, magazines)
- Quiet space
- Communication tools (e.g., whiteboards, flipcharts, intranets)
- Project rooms (e.g., with permanent wall charts, whiteboards)
- Notice boards
- Laboratories and prototype rooms
- Weekly networking meeting
- Mobile computing (e.g., home office, hotel office)

EXAMPLE: SwitchIt Manufacturing has created a detailed activity model for all its manufacturing processes. The purpose of the model is to elucidate these processes and map them against proposed ideas for innovation. A list of the activities is presented in Table 7.3.

Table 7.3   Activity List for SwitchIt Corp.

Activities	
Group	**Title**
A0	Operate SwitchIt Ireland
A1	Manage SwitchIt Ireland
A2	Plan & control manufacturing
A2.1	Plan & control materials
A2.2	Plan & control production
A2.3	Ensure & control quality
A3	Support operations
A3.1	Provide personnel systems
A3.2	Control accounting systems
A3.3	Provide engineering systems
A3.4	Provide information systems

Summary

Innovation begins with the generation of a creative concept that can origi-nate from a problem or idea. Concepts are developed along the action path-way via a series of decision stage gates where suitability is assessed. Projects that successfully flow through the stage gates eventually continue on to be developed into innovations. The creative concepts are identified by employees, customers, or other stakeholders and are recorded and man-aged. One of the key challenges associated with innovation management is to ensure a steady supply of creative concepts to feed the action pathway. Proactive and reactive problems are an opportunity for generating new ideas that can increase revenue, improve efficiency, and lead to innovations. Some ideas do not require significant resources (time or money) and can be implemented immediately. Many more ideas grow to become resource-intensive projects. We now turn our attention to bringing resource-intensive ideas to the next stage of the action pathway as a project.

Activities

This activity requires you to define the key processes or activities of your organization. Create a simple IDEF0 activity list for your organization. Copy Table 7.4 into a spreadsheet and complete the fields.

STRETCH: Create a full graphic model of your organization's key processes and their interrelationships. For assistance on the activity modeling technique, visit www.idef.com.

REFLECTIONS

- What is creativity?
- List and explain five graphic-based tools that can be used for gener-ating ideas.
- Where do ideas come from?
- Explain the mind mapping tool.
- How are failure modes scored in the failure mode effects analysis technique?
- Detail a simple form for capturing critical data for an idea.
- Explain the importance of lead users in generating new ideas.

Table 7.4 Create Activities

Activities	
Group	**Title**

Group: Activity or node number (e.g., A2.1.1)

Title: Activity name in less than 12 words (e.g., "Manage operations")

Managing Projects 8

One of the most important techniques for applying innovation in any organization is project management. Although some good ideas can be implemented immediately as small tasks or "quick wins," the majority take significant resources: time, money, and labor. The words *initiative* and *project* are used interchangeably. They are both nonpermanent goal-centered actions with defined start dates, due dates, and allocated resources. Projects are unique; no two projects will ever be the same. They have leaders and teams. Typically, every organization has a number of projects ongoing at any one time. It is beyond the scope of this chapter to discuss the development of an innovative project in detail and the routines used by project management to achieve this. What is presented in this chapter are the salient features of managing projects in the overall context of managing goals, actions, teams, results, and communities.

LEARNING TARGETS

When you have completed this chapter you will be able to

- Outline the key stages of a project life cycle
- Understand the types of investments necessary for projects
- Explain the importance of risk management in projects
- Construct a simple cost–benefit analysis for a project
- Understand a number of tools for managing projects
- Understand the six stages of the buyer experience life cycle
- Detail a simple form for capturing critical data for a project

Projects

A project is a unique, nonpermanent, goal-centered activity with a predefined life cycle, constrained by cost and resource availability. Projects have a start date and due date. They set objectives that they must strive to achieve within this time window. Organizations undertake numerous projects related to innovation, including the following:

- Installing a new piece of production equipment
- Developing a new technology or science
- Restructuring a department
- Installing a new computer system
- Developing a preventive maintenance program
- Building a new plant
- Developing a new product or service
- Writing a major report

Projects involve changes or planned changes to technology, processes, information systems, and human systems. They are complex undertakings and involve many variables, risks, and assumptions. Ambiguity often exists between a strategic objective and a project, and sometimes the difference can be very small. Consider objectives as broad statements of goals with a fixed time horizon, typically 1 to 5 years. Projects, on the other hand, are the actions the organization undertakes to achieve these goals. Projects have varying start dates and due dates, and with the exception of very large projects, most typically are executed within a strategic planning horizon. All projects have costs associated with the necessary equipment, labor, and services.

A project passes through a number of distinct stages during the course of its life. These stages can be broadly defined as development (which includes conceptualization, design, development, and testing) and realization. Project activities are timed to happen sequentially or in parallel. Managing projects relates to the managing and directing of time, material, labor, and costs to complete a project in an orderly, economical manner and to meet the established objectives of time, costs, and technical or service results. The successful management of a project consists of three major activities: project planning, project scheduling, and project control.

Project Planning

A project can represent a large investment of time, money, and resources. Planning a project effectively can maximize its positive contribution to the company's goals. Good planning is fundamental to the financial consider-ations of a project. It also has organizational implications and is crucial to the scheduling of resources and the control of progress and costs. Delays in implementing projects can be costly and can result in delays in project payback that fundamentally affect the original cost–benefit justification. Delays may also interfere with interfacing systems and can impede the commencement or progress of other projects and the broader scheduling of resources. Effective planning must take into account integration with existing and interfacing systems. A project plan acts as a map to guide people on the project team. Even an imperfect plan is useful because it serves as a pretext to commence work on the project. A complex project can be simplified by drawing up a plan that breaks it up into its constituent parts. The disconnected parts can be examined to see what influence they have on each other, and thus potential conflicts can be anticipated and avoided. Principal elements in the planning of a project are workpackages, tasks, deliverables, gates and milestones, and resources.

WORKPACKAGES

These are a group of tasks, deliverables, and milestones. Projects are often divided into a number of interrelated workpackages to facilitate planning and management of the project implementation. These workpackages typically include the deployment of project objectives. Based on the objectives, the project is divided into tasks. Table 8.1 illustrates one workpackage from a project that has eight workpackages in total. This workpackage outlines the tasks involved in managing the project.

TASKS

These are the principal activities of a project. These activities are con-strained by the project goals and resources. Their results are monitored and deployed against project goals.

DELIVERABLES

These are typically results from the undertaken activities, such as the production of a document or output at a milestone or end of a stage of the project. The responsibility for achievement of a specific deliverable usually is assigned to a single resource or leader in the project (see Table 8.2).

GATES AND MILESTONES

These are points in a project timeline that mark the end of a task or other event, such as a deliverable in the implementation of the project.

Table 8.1   Workpackage Detail

Workpackage
Title: Project Management
Leader: R1 **Number:** WP1 **Start Month:** 0
Objectives To manage the proposed project in a sound manner, addressing such issues as progress monitoring, reporting, reviews, organization, and communication.
O1.1: Effective project management
O1.2: Project control and reporting
O1.3: Liaising with project stakeholders
Description Description of work will be the overall responsibility of the coordinator and is defined to be one workpackage on its own. The work in this workpackage will establish an organizational structure for the project, typically consisting of a project team, a project chair, a technical project manager, and workpackage leaders. Furthermore, to establish a management structure for the project, the workpackage contains the following tasks:
Tasks
T1.1: Establish project, consortium agreement, team portal
T1.2: Manage meetings and goal attainment
T1.3: Undertake progress and cost reporting
Deliverables
D1.1: Project portal
D1.2: Signed consortium agreement
D1.3: Progress report #1
D1.4: Progress report #2
D1.5: Final report
Milestones
M1.1: Completed portal
M1.2: Completed consortium agreement and other contractual documents
M1.3: Submitted progress report #1
M1.4: Submitted progress report #2
M1.5: Submitted progress report #3

Table 8.2   Deliverables

Deliverables		
Deliverable	**Deliverable Title**	**Delivery Month**
D1.1	Project Portal	3
D1.2	Signed Consortium Agreement	3
D1.3	Progress Report #1	12
D1.4	Progress Report #2	24
D1.5	Final Report	36
D2.1	Best Practice in Innovation Management	6
D2.2	Tools for Innovation Management	6
D3.1	Typology of Innovation Processes	12
D3.2	System Model for Innovation	18
D4.1	Concept Explanatory Model	6
D4.2	Design and Implementation Guidelines	12
D5.1	Reference Architecture (Early Freeze)	6
D5.2	Final Architecture and Requirements	21
D6.1	Survey of Standards and Draft Ontology	12
D6.2	Final Ontology for Innovation	24
D7.1	Program Innovation Tool	33
D7.2	Innovation Learning Model	33
D7.3	Distributed Innovation Portal and Tools	33
D8.1	Dissemination Plan and Materials	12
D8.2	Dissemination Report	34
D8.3	Exploitation Plan	34

At milestones or gates, decisions are often made about whether and how the next stage or task of the project is to progress.

RESOURCES

These are teams of personnel, equipment, and finance allocated for the execution of various tasks of the project. Resources often are expressed through individual names or teams and the number of worker-days or worker-months allocated to executing the task (Table 8.3).

The intention of project planning is to smooth the path of an innovation from conceptualization to realization. Planning lays the foundations for coordination and control of a project and can help anticipate trouble and delays. If planning is carried out in a careful, systematic manner, then the implications of conducting each activity of the project can be predicted and potential crises averted. A number of areas in

Table 8.3 Worker-Month Allocation

Code	Workpackages and Tasks	Leader	R1	R2	R3	R4	R5	TOTAL
WP1	**Project Management**	R1						9
T1.1	Establish project, consortium agreement, team portal	R1	3					
T1.2	Manage meetings and goal attainment	R1	3					
T1.3	Undertake progress and cost reporting	R1	3					
WP2	**State of the Art and Best Practice**	R2						18
T2.1	State of the art in intra/interenterprise constellations	R2	2	4				
T2.2	Best practice in supporting computing across industrial sectors	R2	2	4				
T2.3	Learning and innovation for virtual teams	R3	2			4		
WP3	**Innovation Processes**	R4						14
T3.1	Structure and typology in portfolio, program, and project management	R4	2	1	4			
T3.2	Develop system model for distributed innovation management	R4	2	1	4			
WP4	**Learning and Innovation**	R3						18
T4.1	Model of learning and innovation processes	R3		1	2	3	1	
T4.2	Testing and validation of learning model	R3			2	3	1	
T4.3	Development of context-sensitive guidelines	R3			2	3		

(Continued)

Table 8.3 (Continued)

Code	Workpackages and Tasks	Leader	Worker- Months					
			R1	R2	R3	R4	R5	TOTAL
WP5	**Reference Architectures**	R2						18
T5.1	Reference architecture requirements definition	R2	2	4				
T5.2	Developing draft reference architecture	R2	2	4				
T5.3	Validation of reference architecture	R2	2	4				
WP6	**Ontology and Semantics**	R1						10
T6.1	Review of relevant international standards and ontology languages	R1	4		2			
T6.2	Design of an ontology model for innovation	R1	2		2			
WP7	**Prototyping and Design**	R1						31
T7.1	Development of tool for program innovation management	R2	2	2	6		2	
T7.2	Development of innovation learning model	R3		1	2	4		
T7.3	Development of toolset and portals for distributed innovation management	R1	4	2	2	2	2	
WP8	**Dissemination and Exploitation**	R1						12
T8.1	Organize four regional dissemination workshops across Europe	R2	2	2	1	1		
T8.2	Organize a number of focus exploitation workshops	R2	2	2				
T8.3	Develop exploitation plan	R1	2			1	1	
	Total		41	32	29	21	7	130

planning a project can prove problematic. Plans are based on the estimated timing of future events and thus have associated risks. Developing forecasts is difficult, but if they are never attempted, then the solutions cannot be found. Similarly, if no plan exists, then the implementation team will be unable to tell whether they are slipping out of control. Long-range forecasts often are less predictable, and it can be difficult in planning further along the timeline of the project. Detailed planning can be beneficial, but it also introduces an element of inflexibility into the organization's practice. Inflexibility creeps in because of the large amount of hard work needed in planning, resulting in a reluctance to change if flaws are discovered or a more viable alternative arises. A plan is only a working document, and therefore provision should be made in the control systems to regularly question the validity of the plan and update it as necessary.

Project Scheduling

A schedule is the conversion of a project action plan into an operating timetable. It serves as a fundamental basis for monitoring and controlling project activity and, taken together with the plan and budget, is a major tool for the management of projects. There are a number of scheduling techniques, including critical path analysis and Gantt. The Gantt technique is by far the more popular in organizations, and many software tools exist to support the creation and editing of Gantt charts. Displayed against a horizontal time scale, the Gantt chart shows planned and actual progress for a number of tasks. It takes the form of a bar chart, which provides a graphic picture of a schedule. On a Gantt chart the vertical axis indicates the activities to be carried out, and the horizontal axis indicates the timeline. It is an effective and easy way to indicate the current status for each of a set of tasks compared with the planned progress for each item of the set. The Gantt chart can be useful in tracking, dispatching, sequencing, and reallocating resources among tasks. A simple Gantt chart is illustrated in Table 8.4. The activities used in this Gantt chart are the generic activities listed earlier.

Resource planning is another important activity in leading and managing projects. Projects require resources such as people, time, and money. Significant resources mean that more projects can be carried out simultaneously. A balance is needed between the resources available and the activities to be carried out. The basic approach of all scheduling techniques is to form an actual or implied network of activity and event relationships that graphically portrays the sequential relationships between the tasks in a project and also the relationships between different projects. Tasks that must precede or follow other tasks are then clearly identified, in time as well as function.

Table 8.4   Gantt Chart

Code	Workpackages and Tasks	Leader	Year 1				Year 2			
			Y1Q1	Y1Q2	Y1Q3	Y1Q4	Y2Q1	Y2Q2	Y2Q3	Y2Q4
WP1	**Project Management**	R1								
T1.1	Establish project, consortium agreement, team portal	R1	■							
T1.2	Manage meetings and goal attainment	R1	■	■	■	■			■	■
T1.3	Undertake progress and cost reporting	R1		■				■		■
WP2	**SOTA & Best Practice**	R2								
T2.1	State of the art in intra/interenterprise constellations	R2	■						■	
T2.2	Best practice in supporting systems across industrial sectors	R2		■						■
T2.3	Learning and innovation for virtual teams	R3	■							■
WP3	**Innovation Processes**	R4								
T3.1	Structure and typology in portfolio, program, and project management	R4	■		■				■	
T3.2	Develop system model for distributed innovation management	R4								■
WP4	**Learning and Innovation**	R3								
T4.1	Model of learning and innovation processes	R3	■							
T4.2	Testing and validation of learning model	R3		■						
T4.3	Development of context-sensitive guidelines	R3			■					

(Continued)

Table 8.4 (Continued)

Code	Workpackages and Tasks	Leader	Year 1				Year 2			
			Y1Q1	Y1Q2	Y1Q3	Y1Q4	Y2Q1	Y2Q2	Y2Q3	Y2Q4
WP5	**Reference Architectures**	R2								
T5.1	Reference architecture requirements definition	R2		■						
T5.2	Developing draft reference architecture	R2						■		
T5.3	Validation of reference architecture	R2				■				
WP6	**Ontology & Semantics**	R1								
T6.1	Review of relevant international standards and ontology languages	R1		■						
T6.2	Design of an ontology model for innovation	R1					■			
WP7	**Prototyping and Design**	R1								
T7.1	Development of tool for program innovation management	R2					■	■	■	
T7.2	Development of innovation learning model	R3					■	■	■	
T7.3	Development of toolset and portals for distributed innovation management	R1					■	■	■	
WP8	**Dissemination and Exploitation**	R1								
T8.1	Organize four regional dissemination workshops across Europe	R2		■				■		
T8.2	Organize a number of focus exploitation workshops	R2			■				■	
T8.3	Develop exploitation plan	R1								

Project Control

Once a plan has become operational, control of the project is necessary to measure progress, identify deviations, and take corrective action when needed. Control involves analyzing the situation, deciding what to do, and then doing it. Control is concerned with getting the project back on course if reporting reveals lapses in the project's progress. Control may be concerned with time schedules, the completion date for an activity, or milestones and is important in deciding whether the current plan is realistic or attainable. Realigning the plan may entail shifting the milestone date, setting less ambitious targets, applying additional resources, or redistributing the workload. Control is about acknowledging problems and initiating measures to tackle them. The main problem with projects that are off schedule is that although people have identified the problem, they often do not have the power to do something about it. The success of a project is based not just on its on-time completion but also on parameters such as achieving the planned costs and meeting the project's defined objectives.

Project plans are not made in stone and are subject to changes and reviews throughout the project life cycle. Control systems imply a certain amount of inflexibility, but it is incorrect to think that controlling implementation according to the plan guarantees success. On the contrary, often, the more high performing the team, the more it wants to deviate from the plan and improve the specification as it goes along. In practice a good project team and an effective leader constantly modify the project's plans as it develops. They check the modifications with the stakeholders and the external environment, identify any problems, replan where necessary, and renegotiate resources and support if needed. Planning review systems offer greater control of projects and help to move them forward. Planning is an iterative process yielding better plans from the repetitive process of improvement. Continuous cycles of planning, implementation, and reviewing take place throughout the project. The process becomes of greater importance to project success as the uncertainty and risk attached to the innovation increase.

Quantitative Benefits

Most projects that consume significant resources must be justified in terms of costs and benefits. Both quantitative and qualitative techniques can be used to assess the benefits of a project. The core benefit of quantitative techniques is that they encourage the organization to attempt to specify the benefits and costs associated with a particular project innovation, even if these figures are just estimates. This gives management better information about crucial project decisions. Three popular quantitative techniques that can be used to assess project suitability are payback, return on investment (ROI), and net present value (NPV).

PAYBACK

The payback technique is the simplest way to determine the financial attractiveness of a project. It involves three variables: initial cost, recurrent cost, and benefit (annual revenues, cost avoidance, or cost savings). The payback ratio is calculated as follows:

$$\text{Initial cost}/(\text{Annual benefits} - \text{Annual recurrent cost})$$

The ratio provides an indication of the number of years the investment will take to achieve payback. Individual organizations have their own standards for payback periods below which a project may be considered meritorious. Many organizations will not accept a payback period of greater than 3 years. The major advantage with the payback technique is that it is simple to implement and easy to understand.

EXAMPLE: A machine costs $85,000 and generates revenues of $55,000 per year for 7 years. It costs $30,000 to operate the machine. At the end of year 7 the machine is scrapped. Determine the payback period.

$$\text{Benefit} = 55{,}000$$

$$\text{Recurrent cost} = 30{,}000$$

$$\text{Initial cost} = 85{,}000$$

$$\text{Payback} = 85{,}000/(55{,}000 - 30{,}000)$$

$$\text{Payback} = 3.4 \text{ years}$$

This project will pay for itself in 3.4 years. After this time the project will be a net contributor to overall revenue.

RETURN ON INVESTMENT

This technique (also known as rate of return) compares the money earned from the investment against the total investment. The advantage of this technique is that each individual project has its own return on investment percentage, which enables the comparison of projects based on their percentage return. The disadvantages of this approach are that it does not take into account the time value of money or difficulty in determining realistic future values. Another disadvantage is the difficult way in which the percentage return is calculated.

NET PRESENT VALUE

The NPV method discounts all costs and revenues (current and future) associated with a project to a current-day value. If the aggregate value of the calculation is positive, then the project return exceeds the organization's required rate of return. If the aggregate value is negative, then the investment does not meet criteria. For example, an organization may have decided on a required discount rate for all investments of 8%. Any project that results in a percentage lower than this would be deemed unattractive. The benefit of this technique is that it incorporates the time value of money into the calculation and encourages the organization to question the timing of costs and revenue streams. A disadvantage is the difficulty in determining values across the future time scale. Another disadvantage is that the technique requires the use of difficult formulas.

A disadvantage with all quantitative techniques in project assessment is that the figures used often are estimates at best and can be incorrect. Using these techniques gives only part of the picture of the value of the project, but because they are mathematical, people sometimes take the results as absolute. Some meritorious projects (e.g., installing a new computer network) may have a negative result because it can be difficult to directly associate revenue streams with the project. Despite negative quantitative results, this project may be a competitive necessity for the organization. Organizations often need to incorporate both qualitative and quantitative techniques in project assessment to ensure appropriate balance.

Qualitative Benefits

Some projects, such as providing new ambulance services for a hospital, are not easily analyzed in terms of quantitative benefits such as payback. Also, given the obvious risks associated with using quantitative methods, many organizations choose to assess a project's contribution using qualitative criteria. Qualitative techniques are subjective in that they rely on a person's opinion of a project's fit in the organization in terms of a number of specific criteria. Many organizations choose to use qualitative techniques in a team-based environment in order to elicit the full range of views about the value of the project and avoid individual bias. Management can score the project relative to single or multiple criteria and decide on its suitability. The following are some criteria organizations use to assess the suitability of their innovation projects:

- Fit with organizational goals and objectives
- Fit with competitive necessity for sustainability

- Fit with existing product or service range
- Fit with available resources
- Fit with existing competencies
- Fit with desired future competencies
- Fit relative to competitor direction
- Fit relative to risk quotient
- Fit with other innovation projects, ongoing or planned

The technique can simply assign values from, say, 1 to 5 on each of these criteria. It may also assign a weight to each criterion because some criteria may be more important than others. At the end of the analysis a score is determined for each project. When qualitative and quantitative values are combined, people can gain greater insights into the associated costs and relative benefits of a project. This helps organizations decide whether a project should be pursued and, if so, how it can be managed to successful completion. A more detailed discussion of qualitative scoring is provided in Chapter 10.

Risk Management

Projects are future-oriented unique events and therefore involve risk. Risk is the combination of three components: an event in the future or a change that may take place, the probability that this event may occur, and the impact that this event will have. Risk can therefore be defined as a function of these components:

$$\text{Risk} = f(\text{event, probability, impact})$$

Managing project risk can include identifying, analyzing, and responding to project risk. It includes maximizing the results of positive events and minimizing the consequences of adverse events. To do this, the project leader relies on his or her own judgment, experience of similar situations, and tools specifically developed for such a situation. The final decision about a particular course of action may be linked to the tolerance the project leader (or organization) has for risk. Risk management is a proactive rather than a reactive exercise that entails planning and anticipation rather than reaction.

The project leader's tolerance for risk influences the level of risk exposure of the project. Three commonly used classifications for project leaders are risk averter, neutral risk taker, and risk seeker. A risk averter is someone who avoids risk more and more as the money at stake increases. A neutral risk taker makes a balanced decision on the risk, calculated from its probabilities and the possible gains or losses from such a decision. A risk seeker is more likely to make the decision if the opportunity benefit is potentially significant. Understanding the project leader's attitude toward risk and the challenges associated with the specific project can give valuable insights into how risk should be managed and the level of control that should be maintained on the project.

Project Innovation

Resource-intensive projects that involve a number of individuals or project teams are smaller innovation funnels within the larger organizational innovation funnel. The key difference is that large project funnels focus on the scope of the individual innovation project, whereas the organization's funnel focuses on managing the portfolio of all innovation actions (of which the large project is just one). Projects can be complex and require careful planning, execution, and control. Often a structure for the knowledge associated with a project has to be put in place in order to capture all the information associated with the project. Project knowledge is important when the project is being planned and executed but also after the project has been completed so that the organization can learn from the project's strengths and mistakes. To describe project innovation, we can apply an adapted version of the innovation funnel. The adaptation is called the project innovation funnel. The four main areas in the project innovation funnel are project goals, project actions, the project team, and project results.

This funnel is depicted in Figure 8.1. Project actions (e.g., workpackages, tasks, deliverables, milestones) flow through the funnel and are restricted by the project goals and teams (human resources and finance). The results of the project actions are measured and compared against the project goals. Continuous monitoring and sharing of project results against project goals can help move a project toward a successful conclusion.

PROJECT GOALS

The project goals guide the innovation project actions in the right direction so as to align the actions with the overall project aims and objectives. The project goals can be grouped into four modules: project statements, project requirements, project objectives, and project indicators.

PROJECT ACTIONS

Project actions are the activities that flow through the funnel and are constrained by the project goals and the project team. The main project actions are the workpackages, and inside the workpackages, the associated tasks, deliverables, and milestones.

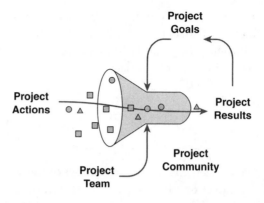

Figure 8.1   Project Innovation Funnel

PROJECT TEAM

The project team consists of the human and financial resources that are available for the project innovation process. The project team also constrains the neck of the project innovation funnel. The experience, skills, and expertise of the project members can limit the innovation that can take place in a project. Training can also influence the effects of human constraints. Similarly, other project resources, such as the capital budget, constrain the scope and level of innovation undertaken in a specific project innovation.

PROJECT RESULTS

The results element represents the outputs of the activities implemented in the project. These results are then analyzed in terms of the level of goal achievement. Based on the feedback drawn from these comparisons, new actions may be initiated or existing actions modified to facilitate better achievement of project goals.

PROJECT COMMUNITY

The final area, community, represents various tools and techniques that can be used to improve communication between project team members, creating a sense of community that fosters better knowledge management. The creation of today's complex systems of products requires a refined mixture of knowledge from diverse disciplines and professions, both internal and external to the organization. This creative cooperation is critical for successful innovation.

Project Tools

Over time a number of tools have proven effective in helping project teams achieve their objectives, shorten development time, and optimize use of resources. These tools include techniques such as concurrent engineering, quality function deployment (QFD), buyer utility mapping, and rapid prototyping. Although many of the techniques have their origins in product development and Japanese quality approaches, they also facilitate the development of product, process, and service innovations. Each of these tools facilitates the development of projects to achieve their objectives, which align the eventual output of the project with the particular market needs.

CONCURRENT ENGINEERING

Concurrent engineering (also called concurrent design) is the simultaneous and coordinated effort of all functional areas engaging in the development process that accelerates the time to market of an innovation. The traditional approach of sequential development, in which the design is developed and then passed to engineering and on to the next relevant function and so on, has proven to be myopic, time-consuming, and erroneous. The approach also increases conflict between related functions because of disagreements over errors and changes. Organizations that adopt concurrent engineering to develop their innovations enhance communication throughout project development by engaging all relevant stakeholders in decision making as early as possible. This engagement allows activities such as the product design and process design to be undertaken in parallel and for each activity to reflect the others' requirements. Although the team will spend more time completing the concept phase in the concurrent approach than in a similar sequential approach, the greater communication and understanding will significantly reduce the number of design changes that have to be made during the later phases of development. The concurrent engineering approach brings together multifunctional teams in the early phases of the development process and contributes to a collaborative rather than confrontational environment within organizations. Trust and understanding develop between the functions, and team members resolve difficulties without assigning blame. The concurrent approach also increases the likelihood that the eventual output of the development process will meet customer needs because marketing can be involved in the process throughout. As concurrent engineering reduces the number of revisions that must be undertaken, it reduces the overall development time significantly (in some cases by more than 40%). Potential innovations make it to the market faster than they would under the traditional sequential process and gain the competitive advantage of being first to market, which can increase the potential success of the innovation. Illustrating the need for speed and flexibility in the new product development process, Takeuchi and Nonaka (1986) emphasize six characteristics that combine to support innovation. These six organizational characteristics are built-in instability, self-organizing project teams, overlapping development phases, multilearning, subtle control, and organizational transfer of learning. Together these characteristics can act as a stimulus for driving innovation throughout the products, services, and processes of established, rigid organizations.

QUALITY FUNCTION DEPLOYMENT

QFD is an approach developed by Japanese industry to enhance the process of translating customer requirements into a product design by considering the voice of the customer at each stage of the development project (Hauser & Clausing, 1988). The main features of QFD are as follows:

- Meeting market needs by defining actual customer requirements
- Applying a multidisciplinary team approach
- Using a matrix approach for documenting relationships

The process begins by identifying the important customer requirements, which are typically compiled in collaboration with the marketing department. The relative priority of each of these requirements (from the customer perspective) is scored. The relationship of the various requirements is then mapped to specific product or service characteristics to indicate the level of correlation. Obviously, customer requirements that are not reflected in product characteristics can result in the addition of new product or service features, and similarly product characteristics that do not relate to requirements may be removed from the eventual product. The QFD approach allows an organization to assess the strength of its ability to meet customer requirements relative to other competitors in the market. This information can highlight weaknesses in the potential innovation and result in revisions to the eventual design specification. The QFD technique ensures that customer requirements are incorporated at all stages of development. This will increase the likelihood that the eventual offering will meet market needs and will be adopted by the customer base. The QFD technique may be applied to the development of any new product, process, or service using several steps whose results are presented in a unique arrangement of matrices and tables known as the House of Quality. This is illustrated in Figure 8.2.

BUYER UTILITY MAPPING

This technique is used to assess the relative appropriateness of a project using a two-dimensional matrix with six utility levers on the *y*-axis and six

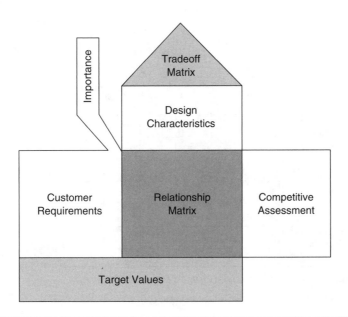

Figure 8.2 House of Quality

stages of buyer experience on the *x*-axis (Kim & Mauborgne, 2000; see Figure 8.3). The approach suggests that every customer measures the utility of a product, process, or service according to the criteria on the map. The six utility levers are productivity, simplicity, convenience, risk, fun and image, and environmental friendliness. If the relative utility of a concept or product is better than a competitor's, then the idea is a good one. An alternative application of the buyer utility map is to use it to compare the potential innovation relative to customer requirements rather than a competitor's product.

EXAMPLE: A low-cost airline recently completed a buyer utility map for its innovative service to airline passengers. They later contrasted their map with that of a competitor's buyer utility map. The map is illustrated in Figure 8.3.

During the purchasing stage, use of the Internet contributes toward lower cost (i.e., collective productivity), convenience, lower risk (because the customer interacts directly with the airline), and fun and image (because customers often boast to friends about the price of their tickets). Simplicity is not seen as an advantage because interacting with a Web site is not as simple as interacting with a travel agent. During the delivery of tickets, the use of the Internet as a delivery mechanism provides numerous advantages. During the use of the service, customers view the no-frills model as advantageous in almost all utility levers. Finally, in the supplements phase, activities such as changing dates and repurchase due to missed flights again lead to a more productive service because new tickets purchased on the spot are seen to be inexpensive. The maintenance and disposal stages are not relevant for this service.

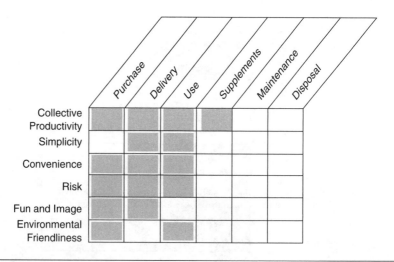

Figure 8.3   Buyer Utility Map for a Low-Cost Airline

SOURCE: Adapted from Kim and Mauborgne (2000).

RAPID PROTOTYPING

This involves the physical production of components of the product or services. Despite the information provided by computerized systems and simulations, a physical or computer-based prototype often provides richer information. The development team can develop prototypes of varying levels of complexity in order to test and analyze certain features. Physical prototypes are often used as a means of validating the development direction with the market because the customer can interact better with the physical prototype than with a computer model. Physical prototypes can show potential users factors such as touch, feel, and weight of the prospective product that cannot be communicated by concept or computer modeling alone. The downside is that physical prototypes often are much more costly and time-consuming to develop than their virtual counterparts. During the development of a potential innovation, both computer and physical prototypes may be developed depending on the type of test and validation needed. The purpose of this is to provide the team with better information, which will enable them to make decisions about the advancement of the project and to avoid costly mistakes in later stages of the innovation process.

Summary

Project management is an important part of managing innovative actions. In this chapter we looked at a number of features of project management that are important in the overall context of managing goals, actions, teams, results, and communities. Project management is essentially about managing schedules and resources. Each innovation project goes through a life cycle that can involve compromise between the constraints of goal achievement, time, and cost. Projects are assessed for continued suitability at each stage of the action pathway, using both quantitative and qualitative techniques. Care must be taken that the organization manages the entire process from concept through commercialization if the innovation is to be successful. A number of tools can facilitate the development of an innovation project to a successful conclusion. A special case of innovation is new product development that includes additional stages such as intellectual property protection and commercialization. When commercializing the new products, the organization must manage its transition across all the market segments of the life cycle in order to ensure that maximum benefit is recouped from the investment. The next chapter will look at a number of new product development issues.

Activities

This activity requires you to create details for one long-term project that will result in innovation for your organization. Focus on creating a list of workpackages and tasks for the project, together with a set of suitable milestones. In addition to defining a project leader, define the support team that will facilitate the implementation of the particular project. Copy Table 8.5 into a spreadsheet and complete the fields. Add new columns and rows as needed.

STRETCH: Other elements of this activity may include creating a list of deliverables and creating a network chart to show critical paths where appropriate. Critical information on each workpackage may also be outlined on a separate page.

REFLECTIONS

- Outline the key stages of a project life cycle.
- Explain the issues around planning, scheduling, and controlling a project.
- Explain the difference between workpackages, tasks, and deliverables.
- Construct a simple cost–benefit analysis for a project.
- Explain the following expression: Risk = f(event, probability, impact).
- What are the six stages of the buyer experience life cycle?
- Detail a simple form for capturing critical data for a project workpackage.

Table 8.5 Project Tasks

Project:

Code	Title	Resp.	Worker-Months						Year 1				Year 2			
			R1	R2	R3	R4	R5	TOTAL	Y1Q1	Y1Q2	Y1Q3	Y1Q4	Y2Q1	Y2Q2	Y2Q3	Y2Q4
WP1																
T1.1																
T1.2																
T1.3																
WP2																
T2.1																
T2.2																
T2.3																
WP3																
T3.1																
T3.2																
WP4																
T4.1																
T4.2																
T4.3																
Total			0	0	0	0	0	0								

Title: Title of the work package or task

Resp.: Individual responsible for leading the project

Man-months: Number of man-months allocated to each work package and task

R1: First resource (e.g., Individual or Team or Organization)

Total: Total number of man-months for each workpackage or task

178

Developing Products 9

Radical projects that lead to new products regularly make the headlines as examples of good innovation. Technological innovation (innovation that results from the adoption of new technologies) is regularly cited in various news media, journals, and books. Although these kinds of projects are often entertaining and certainly illustrative of innovation, they represent only a small proportion of the innovation taking place in organizations. Nonetheless, many useful lessons can be gained from a study of new product development or product innovation in commercial organizations. The special case of product innovation deserves attention, which we provide in this chapter. New product development is about creating new products that add value to customers, fit with their expectations, and can be achieved cost-effectively. This chapter presents a number of core concepts around new product development in commercial organizations.

LEARNING TARGETS

When you have completed this chapter you will be able to

- Outline the key stages in the stage gate process for new product development
- Detail project investment issues for new product innovation
- Define a number of ways to share investment in product development
- Outline a number of ways of protecting innovations
- Explain the product exploitation process
- Discuss the special role of entrepreneurship in product exploitation

179

New Product Development

The development of a new product can take significant time and resources to bring an initial concept to successful market launch. Although certain electronic products such as the iPod may have been developed in less than a year, the development of other products such as a new drug can take up to 12 years. Time to market can be crucial to the success of an innovation. Being first to market gives the organization an initial monopoly. It also emphasizes the innovative nature of the offering because competing organizations have not yet managed to introduce a similar offering. Being first to market can offer an organization the opportunity for premium price and ease of access to distribution channels. Consequently, delays in bringing an innovation to the market can reduce competitive advantage, reduce market share, and limit the ability to charge premium prices. Similarly, because of the greater likelihood of other competitors in the market, the organization will have to invest more in marketing to differentiate its offering from that of its competitors.

In the technology product sector, anecdotal evidence suggests that a 6-month delay in reaching the market can result in a one-third reduction in profits over the life of the product. Being first to market has obvious advantages, but it can also have associated challenges, such as educating the customer and creating the market (if it did not exist before). Often this can take significant time and marketing funds. The danger is that if the organization has not adequately protected its intellectual property, then others may be able to replicate the innovation once the viability of the market is proven. Certain organizations adopt a strategy of allowing their competitors to enter new markets first; they will enter these markets themselves only when the commercial opportunity is proven. Organizations that adopt this type of strategy often have competitive advantages such as size, brand name, or economies of scale that they can leverage in order to enter the market and counteract the benefits of the first-to-market organizations. Irrespective of the strategy the organization eventually adopts, time-to-market concerns must be factored into the development planning of product innovation.

Stage Gate Process

The stage gate process is a powerful tool for managing the new product development process (Figure 9.1; Cooper, 2000). Cooper describes the stages by which products flow from concept to market as a generic stage gate process consisting of sequential stages. These stages include the ideation stage, preliminary investigation stage, detailed investigation stage, development stage, testing and validation stage, and full production and launch commercialization stage. Cooper also includes a postimplementation

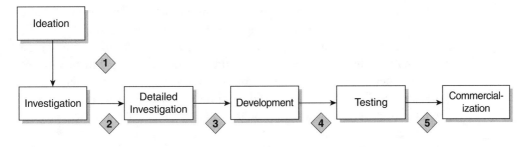

Figure 9.1 Stage Gate Process
SOURCE: Adapted from Cooper (2000).

review phase at the end of the process to allow feedback and learning to occur across the organization.

IDEATION AND PRELIMINARY INVESTIGATION STAGES

Cooper's stages for new product development begin with an initial ideation stage in which the ideas are generated and initially screened for suitability. The concept passes to the preliminary investigation stage, where a quick and inexpensive investigation is undertaken. After this stage, the concept passes through a second stage gate, where it is evaluated relative to other potential ideas that have been proposed. If deemed inappropriate, the concept is abandoned; otherwise, it continues to the next stage of the process.

DETAILED INVESTIGATION STAGE

At this stage, a detailed investigation of the concept is undertaken in order to build the business case for the prospective innovation. This stage studies the concept in much more detail and tests the market and technological potential of the concept. It is in this stage that significant information is compiled about the concept in order to help managers decide whether the idea should flow to the development stage or be abandoned. The richer and more reliable the information developed about the future for the potential innovation, the better the quality of management decisions made at this stage gate.

DEVELOPMENT STAGE

The next stage of Cooper's process views the concept undergoing development. This is a highly expensive stage of the process as the product begins to become tangible. Challenges such as technological feasibility are addressed in this stage through use of the knowledge competence of the organization and through ongoing experimentation and testing. This stage attempts to align the aspirations of the concept with the technological feasibility of the state of the art. The duration, expense, and difficulty of this stage depend on the relative ambition of the concept. In progressing a potential innovation through the development phase, the team can also make related discoveries and breakthroughs as a byproduct of the process. These discoveries often find their way to the ideation phase of the process and become the creative seed for other innovation projects. Although successful ideas pass through this stage having conquered their technological challenges, many concepts are abandoned because the project scope or the skills needed are beyond the capabilities of the organization at the time. Depending on the outcome of the development stage, managers must decide whether the concept should be abandoned, allowed to continue in development, or progressed to the next stage. This decision is influenced by the progress made during development of the project, by the associated costs (to date and future), and by the attractiveness of the concept in terms of market opportunity, given current information. If management thinks that the concept is likely to progress to the next stage of the process, then commercialization plans relating to production, marketing, and distribution of the innovation are developed in parallel with technological development.

TESTING AND VALIDATION STAGE

This stage involves tests of prototypes relative to desired performance, market expectations, and production systems. Positive outcomes can result in the product flowing on to the final stage of the process, but unsatisfactory results can often cause the concept to return to the development stage for rectification of problems. Sometimes (as is the case in the pharmaceutical development process) the inadequacies highlighted during the testing and validation stage can be so significant that they result in the abandonment of the concept altogether. The difficulty managers face at this stage gate is to decide where further investment in the concept would be prudent given the substantial investment already made. This decision is influenced by the challenge of the inadequacies highlighted and the potential opportunity offered in bringing the concept to market. If the organization allows the potential innovation to continue along the action pathway, it will reach the full production and market launch stage. This stage can also be called the commercialization stage.

COMMERCIALIZATION STAGE

During this stage the production plan is implemented to allow increases in the scale of production to supply the market. If the development stage has not taken adequate account of production and market needs during the determination of the final design, then difficulties can be encountered. This stage ensures that full production is achievable, given the quality, cost, and time parameters. Similarly, the marketing plan is implemented so as to launch the potential innovation to the market and ensure a satisfactory response. Here issues such as market education, positioning, and after-sales support are put in place. This stage also defines the proposed life cycle of the potential innovation and how it will be monitored so as to allow for improvement or corrective action if necessary. The result of a well-managed stage gate process is a successful innovation that achieves the performance parameters set by the organization and, most important, adds value to the customer. Cooper includes a postimplementation review phase as a conclusion to his stage gate process. The purpose of this phase is to allow the project team to take a holistic overview of the implementation of their particular product innovation. This allows feedback of information that may be pertinent to any future product versions and also provides the team and organization with the opportunity to reflect and learn from the project.

STAGE REVIEW

At each stage gate, the project team and senior management review the progress of the project during the previous stage. This review results in a decision about the fate of the project. If the outcome of the review is satisfactory, then it is allowed to proceed to the next stage of the process. Otherwise, the decision can be made about whether the project should return to the previous phase for further development or be abandoned altogether. The decision to abandon the project completely can be made in scenarios where initial development and testing have highlighted unforeseen challenges that change the attractiveness of the particular project. Similarly, the external environment can change (or not develop as planned) during the project implementation, and therefore the market need is not as evident as initially thought. As the particular project is developed through the various phases, the stage gate reviews inform the organization about progress toward its goals. This information allows the organization's senior management to intervene if necessary and influence the project performance by increasing the resources available to the team or altering the skill mix and experience of the team. The organization can assess the progress of the project by using both quantitative and qualitative analysis techniques.

Product Funding

The innovation process can be a long and costly road that the organization must follow in order to produce the innovations necessary for it to survive and grow in the current competitive environment. The majority of innovations are developed over a number of years. In many instances this produces a cash flow gap that must be addressed (Smith, 2006). Even large-scale multinational organizations can encounter difficulties in funding the cash flow gap over the duration associated with the innovation's development. One estimate of the development cost of a new medicine from the laboratory bench to market is more than $400 million (Light & Lexchin, 2003). Other sector experts estimate costs to be closer to $1 billion and the duration from discovery to market to be more than 12 years. Consequently, modern organizations are seeking ways to address the financial constraints so that they can engage in innovation. Organizational management is being forced to examine both internal and external sources of funding for innovation.

Organizations prefer to rely on internal funding of their innovations because sole ownership remains within the organization, and the associated costs of finance are reduced for this reason. The main source of internal funding comes from company reserves built up over the years of profit and represented as an asset on the balance sheet. The organization plows its reserves of retained profits into the development of the innovation so that the revenue generated from sales of the project will increase future reserves, allowing the cycle to repeat itself. On occasions the organization will lack the reserves necessary to fund the development, because of either the financial situation of the organization or the magnitude of the innovation. Under these conditions, it is forced to look for external funding or forgo the opportunity for development. External funding can originate from a number of sources, including government grants, bank loans, business angels, venture capitalists, and strategic alliances (Figure 9.2).

Each of these sources of funding is applicable at various stages of the innovation process, but there are also associated costs. The main cost is in terms of the ownership that the organization must relinquish as it continues down the list of funding options. Organizations choose to use the various external funding sources when they think the revenues from future innovation, despite the associated costs of funding, are too attractive to forgo for lack of resources.

GOVERNMENT GRANTS

National governments want to encourage organizations to engage in innovation in order to ensure the ability of organizations to provide future employment and tax revenue for the national economy. Many national economic policies have been built on the platform of the knowledge economy, and innovative capability is essential to generate wealth from the knowledge base of a nation. Therefore, governments occasionally provide

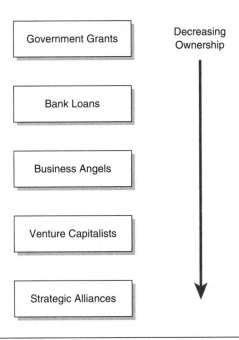

Figure 9.2   Types of Funding

grants and public contracts to stimulate innovation activity in a particular industrial sector, such as biotechnology or information and communication technology. These public funds can be a valuable resource for supporting organizations during the early stages of the process, such as the discovery phase, where other funding sources may be less inclined to invest because of the risk levels. The terms associated with these grants vary significantly, and under certain circumstances (usually in relation to commercialization grants) government bodies can insist on taking an ownership share in the innovation. One of the key criticisms leveled by industry at this mode of innovation funding is the associated bureaucracy, which often deters organizations from using this source of funds.

BANK LOANS

Organizations often use banks as a source of funding for the development of innovation. The primary type of bank funding is debt funding, in which certain funds are lent to the organization, to be repaid over a period of time, together with an associated interest repayment. The advantage for the organization is that sole ownership is retained in the organization, and it is able to bridge the funding gap until profits from this year's turnover can be invested internally. Though attractive to organizations as a source of funding, many banks are reluctant to provide funds for innovation-related projects because of the associated risk, and they engage only with organizations that can provide adequate collateral for reassurance. Therefore, this source of innovative funding can be beyond the reach of many small and medium-size enterprises unless owners are willing to use private assets as collateral.

BUSINESS ANGELS

These are sources of funding used by private organizations when they are unable to acquire or have already expended the sources of funding already discussed. Organizations leave these sources of funding until the later stages because of the share of ownership the organization must relinquish in order to obtain this funding. Business angels are usually wealthy private investors who are seeking an attractive return on their capital. These people are often entrepreneurs who have cashed out of their own businesses and are seeking both a return on investment and an outlet for their entrepreneurial interest. Although many business angels adopt a passive role in the company, allowing management to continue operating normally, others require a more active role in which they influence the strategic direction of the organization as well as provide funding. This additional input of business insight and experience can be advantageous but can also have a disruptive effect on the management team and on innovation development. This mode of funding usually is used by startup organizations that lack the internal reserves and reputation to develop the innovation on their own.

VENTURE CAPITALISTS

These are also investors seeking a high return on their investment, but they are also typically professionals who can coordinate large investment funds. Therefore, the sums of money at their disposal are usually larger than those of a business angel, and they are also interested in what exit strategies are available so they can cash out their investment within a definite time window. Venture capitalists invest the money of others, and these people expect their funds to be returned with significant growth after a certain period of time. These investments can be recouped through the sale of shares in the company, or, in the case of a non–publicly traded company, a trade sale can be used. Acquiring funds from business angels or venture capitalists involves a partial loss of ownership. Although traditionally this mode of funding has been used primarily by entrepreneurial startup organizations, in recent years more established organizations have begun to use this source of funding for their innovative initiatives.

STRATEGIC ALLIANCES

The final source of funding for innovations is that of strategic alliances, and these are becoming a more common way to finance the development of innovations. Whereas the previous external sources of funding are used primarily by startup and small and medium-size organizations that lack internal reserves, this mode of financing can be used by large organizations. Organizations occasionally choose to partner with other independent organizations in order to develop a prospective innovation. This allows the organizations to spread the cost of the innovation across the partnering organizations, thus reducing the risk exposure of any one organization. The partner organizations agree on a shared ownership of the innovation and any related intellectual property that arises through the development process. Details about the input of resources (e.g., finance and human resources), ownership and division of revenue streams, and management of the alliance are normally agreed on in advance and defined in a legal contract. The selection of a partner organization is influenced by the ability to co-fund and equally by the ability to contribute to the development of the innovation through technology and market knowledge. This collective approach to innovation development has been driven by both the increased cost and the technological complexity of developing innovations for the modern organization. This mode of co-development, where innovation is distributed across a network of organizations, will become more common as individual organizations struggle to obtain the internal resources necessary to develop their own innovations.

Protecting New Products

Protecting the investment that the organization makes in the research and development of innovation is advantageous for an organization because it can prevent other competitors from exploiting their knowledge. Although there are many mechanisms for protecting intellectual property, the four main ones are patents, copyright, design rights, and trademarks.

PATENTS

Patents are a legal mechanism created by the state that provide the patentee with exclusive rights for the exploitation of the development in return for full disclosure. Not every innovation is patentable. In order to acquire a patent the innovation must be novel, must involve an inventive step, and must have a practical application. The protection offered by a patent typically lasts for 20 years. Because there is no global patent, the process can be expensive; one must acquire patents for each country where protection is needed. The process begins at the patent office and ends with the awarding of a patent. Although patent fees vary from country to country, it is not unusual for an organization to spend more than $40,000 to secure patent rights on an invention. Once the patentee has secured the patent, it must pay annual fees to maintain the patent. The benefit of the patent is that the patentee has exclusive rights over its invention and can prevent other organizations from replicating it. If another organization infringes on the organization's patent, the patentee can seek redress through the legal system of that country. If the patentee decides not to commercialize the invention, it can choose to license the intellectual rights to another party as part of a financial deal. Patents are a powerful way of protecting an organization's inventive and innovative output and provide an organization with a degree of confidence in its ability to protect and generate revenue streams from this innovation. Although patents provide protection for inventions resulting from an organization's investment, pursuing patent infringement through the legal system is expensive and time-consuming, especially if the accused is an organization with deep pockets. Awarding patents can also benefit society in that new knowledge is made readily available as a base for future advancement instead of being kept as a trade secret by the discovering organization. Thus, this sharing of knowledge increases the global advancement of research and the plentiful supply of inventions and innovations that result.

COPYRIGHT

Copyright is another approach to protecting the intellectual property of an organization. Copyright is a right that comes into effect through creative effort that results in creative output such as written materials, computer software, artwork, music, and similar materials being protected by law. Copyright assigns ownership to the creator and prevents the work from being copied without authorization. Although a product invention would not be protected under this mechanism, the associated design specifications would be protected by copyright. There is no requirement to register work for copyright; it is automatically assigned when the work is created. Copyright affords the creator the right to be identified as author, to copy or adapt the work, to sell copies of the work, and to prevent derogatory treatment of the work. If someone infringes on the author's copyright by, for example, making unauthorized copies, then the author can seek legal redress through the courts. Copyright provides rights to the authors for their lifetime plus 70 years afterward.

EXAMPLE: Napster became one of the most famous names on the Internet in the 1990s. Napster developed a peer-to-peer file sharing program, which it made available through its Internet site. The software allowed people to swap files with each other quickly and simply. Many of the files people were sharing were music files. Because users made their personal music files (purchased legally in the form of CDs) available to share across the network, other users could access these new music files and save them on their computers. Obviously, this situation pleased neither the record companies nor the recording artists because it decreased demand for new CDs (which included copyright royalties to the recording artists). These parties claimed that Napster was facilitating copyright infringement and succeeded in having the site shut down. Napster has relaunched since then but now uses a format that ensures that copyright royalties are paid.

DESIGN RIGHTS AND TRADEMARKS

These are other forms of intellectual property protection applicable to innovations. Design relates to the appearance of a product, and design rights prevent others from copying this design. Design rights can be either registered or unregistered. Registered design rights provide up to 25 years of exclusive rights to the applicant. One of the best-known protected designs is that of Coca-Cola's contour bottle. Unregistered design rights can be acquired only on original designs and operate in a similar manner to that of copyright. Trademarks are any distinctive mark that uniquely identifies an organization, product, or service. The protection of a trademark can be of the utmost importance to an organization because it encompasses the goodwill of the market toward the organization or product, and it provides protection from other organizations copying goods and damaging the reputation of the trademark.

Commercializing New Products

Innovations fulfill customer needs and generate benefit for the organization. When an organization develops a product or service innovation, the issue of how the innovation will be commercialized and brought to the market is critical to success. Business history is replete with stories of companies that were successful in developing a product or service in the laboratory but were unable to market it successfully. Issues such as compliance with industry standards, compatibility with existing products, production costs, distribution capability, and after-sales service can all affect the attractiveness of a new product or service to the customer. Cooper (2000) identifies this as the fifth stage of his stage gate process, where the organization develops both the production plan and the marketing launch plan for the prospective innovation. Although planning implementation occurs in this phase of the process, it is important to highlight that both production and market plans would have been developed parallel to the design and development of the innovation during the earlier stages of the process. In this way, delays and costly redesigns are minimized across the process, and synergy is achieved between the design of the product, the production capabilities to manufacture it, and the marketing strategy by which it will be sold.

PRODUCTION PLAN

The production plan relates to issues such as manufacture and distribution of the product to the market. Decisions about whether to manufacture in house or outsource can have significant cost implications for the selling price to the market and its attractiveness to the customer. Dyson decided to move production of some of their products from the United Kingdom to lower-cost economies so that they could deliver to the market at an acceptable price. If this decision had not been made, the higher price would have deterred customers from purchasing their innovative products, leading to a downturn in sales. The production plan ensures that the organization is capable of meeting increases in market demand by manufacturing the product in a cost-effective manner and delivering it to the market on time. For example, if a company in the toy industry develops an innovation but fails to get it into the shops for Christmas, then they could forgo up to 60% of annual sales.

MARKET LAUNCH PLAN

The market launch plan addresses issues such as which market segments should be targeted, what price and positioning strategy should be adopted, and what marketing media should be used to educate and communicate with customers. The launch of many new products and services to the market is accompanied by significant marketing campaigns that communicate the advantages and develop the correct brand image for the product. Issues such as the timing of the launch and decisions about rollout of the product across the globe are also addressed in this phase. As products embed in the marketplace, they go through a life cycle of growth to decline. The target customer for the product changes as it progresses through its life cycle. The life cycle of a product can be viewed as consisting of a number of target segments that each have specific needs and requirements. If the innovation is to be a success, then the marketing plan must manage the transition across these customer segments.

Linkages With Marketing

The commercialization phase in new product development is crucial to the success of any innovation but often does not receive the attention it deserves. Many organizations adopt a technological focus with respect to their product innovation and assume the market will embrace the new development. Unfortunately, the market response is rarely as accepting as this, and all successful innovations are a marriage of technological and market development. This marriage ensures that when the innovation is launched, not only is the development technologically ready but so is the marketplace. Marketing in the innovation process has two roles. The first is to inform stages of the innovation process such as ideation, development, and testing of the market's needs and requirements. The second role of marketing relates to the preparation and education of the market for the launch of the innovation.

As part of this market preparation, marketing will decide on issues such as the timing of entry, positioning, and the scale of the entry into the market (Trott, 2005). Similarly, in relation to both product and service innovations, marketing will need to decide on the correct mix of the four *P*s if the innovation is to be accepted by the marketplace. The four *P*s relate to decisions about the desired *product* features; the sales *price* and its position relative to competitor offerings; the *placement,* or how the product will be distributed and retailed to the customer; and, finally, the *promotion* in terms of how the product will be promoted and advertised to encourage its adoption by the market.

Many of these marketing decisions are made well in advance of the realization phase, but they can have a significant impact on the offering's attractiveness to the end customer and its movement across market segments. Marketing must communicate a number of characteristics of the product to the prospective market to enhance its adoption. These include the following (Doz & Hamel, 1998):

- Relative advantage of the new offering over its predecessors and competitors

- Compatibility of the new offering with existing skills, technological platforms, and industrial standards

- Complexity, relating to how easily the offering can be understood and used by the customer

- Trialability, relating to the amount of opportunity that the potential customers have to test the offering, to learn more and reduce their risk perception

- Observability, which reduces risk perception and relates to the degree to which potential customers can see other consumers use and benefit from the offering

The stronger the case marketing can make to customers in relation to each of these five aspects of the new offering, the greater the likelihood that it will be adopted by the market and become a commercially successful innovation.

Adoption of New Products

When a radical new product is launched into the marketplace, it can be adopted by various customer segments during its life cycle. Rogers (1983) lists five such segments: innovators, early adopters, early majority, late majority, and laggards (see Figure 9.3).

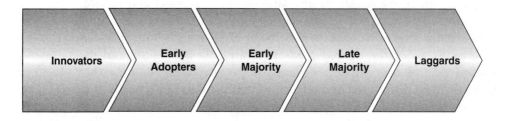

Figure 9.3   Stages of Adoption

INNOVATORS AND EARLY ADOPTERS

Innovators and early adopters value the technological advancement of the product and its advancement over the previous generation. Often the opinions of these customers are sought concerning early prototypes during the product development phase. Unfortunately, these customers do not purchase in large quantities, but they can provide rich feedback on product performance and ways in which the product can be enhanced. The function of marketing to this type of customer is to ensure that key opinion makers have access to the new product and that their feedback is communicated back to the development team.

EARLY MAJORITY

As the product establishes itself in the market, it is used by a larger group of early majority customers, and sales can grow exponentially. The difficulty is that this customer group often has different demands than those of the earlier customer groups. One source suggests that in certain cases (especially with technology innovations), the life cycle is not smooth but instead has a chasm between the early adopters and early majority customer segments or customer groups (Moore, 1999). Only when the organization provides the early majority market with a product to match their requirements will it succeed in crossing this chasm. The early majority is interested in the function of the overall product and its contribution to their utility level rather than its technological superiority. Issues such as quality, reliability, value, ease of use, and after-sales service are some of the factors important to the early majority customer group.

LATE MAJORITY AND LAGGARDS

If the innovation is adopted by the early majority, then it will progress across the late majority and laggards customer groups. At this stage, cost per unit can decrease as market exposure to the concept or production runs increase.

The key challenge in commercializing an innovation is to manage its adoption by the early majority so that significant sales can be achieved and investment recouped. Ongoing market research throughout the process will have determined the market requirements and the value of the innovation to various customer groups. It is this value, rather than any cost-based calculation, that determines the price the market will sustain. It is the function of the marketing plan to communicate to customers how the innovation fulfills their requirements and thus influence the value perception of customers. Once the product is launched on the market, periodic review is essential to elicit customer feedback and assess where modifications can be made to increase market satisfaction. Although commercialization is most easily applied to product innovations, many of its elements are also applicable to service and process innovations. Issues such as installation and commission of the new process, training and development of employees regarding operation, and periodic review of performance are all important elements in the management of successful process and service innovations.

Entrepreneurship

This book is focused primarily on innovation by an established organization that typically has a portfolio of products, processes, and services. However, there are many examples of innovations initiated by individuals that have created new organizations around what was initially a single idea.

When Schumpeter (1942) described innovation as "the waves of creative destruction" that generated economic growth, he viewed the entrepreneur as the key driver necessary to overcome resistance from the static establishment and make the innovation happen. One can see how almost every major corporation begins life with an entrepreneur at the helm. For example, Bill Gates began with one idea, around which he created the organization known to us now as Microsoft. Entrepreneurs observe opportunities and, with great determination, exploit these opportunities to gain competitive advantage in the face of competition. Entrepreneurs find a new way to attract customers by building an organization that can deliver value through productivity, fun and image, environmental friendliness, or convenience. The ability to identify and develop the idea is matched by the ability to build an organization around it. Entrepreneurs and their activities have clearly demonstrated that they can add significant wealth to an economy in a very short time. There are a number of core lessons to be learned around building an effective entrepreneurial organization. Earlier we looked at topics such as idea generation and business financing; three more important topics are entrepreneurial zeal, business planning, and exit strategy.

ENTREPRENEURIAL ZEAL

Entrepreneurs have a special skill for identifying opportunities, but another important ability is the zeal in problem solving and in building up an organization that is capable of continuous learning, at least until its objective is reached. Entrepreneurs tend to be overachievers with an eye for opportunity and have a high tolerance for risk. They are self-starters, unwilling to wait for others to act, and spend significant amounts of time pursuing their dream. They are generally good decision makers who sometimes act on instinct, although perseverance, tenacity, and willingness to learn from mistakes can sometimes be substituted for this valuable attribute. Entrepreneurs take risks with their own resources, often at the expense of such mundane requirements as a place to live or security for the future. Entrepreneurs display exceptional commitment, often against terrible odds, in getting their organization (and its innovation) off the ground. They are good negotiators, display inordinate optimism at startup, and are prepared to take on much more powerful and successful organizations at their own game.

BUSINESS PLANNING

Many aspects of business planning have been dealt with elsewhere, such as defining goals and managing actions. Here we will look at the structural aspects of creating a business plan. A business plan is similar to an innovation plan except that the business plan is intended to create a new organization and typically involves development of one idea, whereas an innovation plan applies to established organizations and deals with many ideas simultaneously. The purpose of a business plan is often simple: to receive funding or support to begin growing an organization that will add value to customers by exploiting an idea. The main audiences for a business plan are support agencies, funding agencies, banks, venture capitalists, and so on. Business plans can vary in content but generally stick to a predefined format that will allow the plan to be evaluated more effectively. A typical structure includes the executive summary, product or service offering, competitors, management team, operation strategy, market research and marketing strategy, sales projections and finance strategy, and résumés.

EXIT STRATEGY

Most entrepreneurs (and their funding partners) are mindful of an exit strategy. There are a number of reasons to prepare for an exit from the business that has been developed: The business may best be served by merging or dissolving into another organization; the owner may want to cash out of the business to release the accumulated value or; the business may have reached the end of the line and stopped growing, and it may lack the potential for future growth. Some entrepreneurs simply seek more challenges and want to exit from one successful organization in order to invest the capital they have realized in a new venture, starting the whole entrepreneurial process all over again. Entrepreneurs can exit their organization by launching an initial public offering, selling out to a larger organization, or selling to another partner or senior management within the organization.

Summary

New product development is a special type of innovation in commercial organizations that can lead to new technological innovations and even to the formation of new organizations. The new product development process has been the topic of many books and is clearly a very important growth inducer for large national economies. New product innovation can consume very large amounts of capital, so organizations typically look to funding agencies for risk sharing; these organizations will also share the potential benefits. The very high value placed on new products makes it necessary to protect them from potential competitors. Commercialization of new products is the final stage in a long process of realizing the benefits of the new product innovation process. Often commercialization occurs through the process of entrepreneurship, where a new organization is often grown around a new idea.

Activities

This activity requires you to create a list of seven fictitious projects for your organization. Some of these may come from previously defined ideas. Some may also have stemmed from the problems you created earlier. Try to include one or two radical and high-risk projects that will have ambitious objectives, consume substantial resources, and span a significant time frame. Try to ensure that the majority of the projects have shorter durations than the planning period (e.g., 3 years) and cover all innovation types (product, process, and service) appropriate to your organization. Copy Table 9.1 into a spreadsheet and complete the fields. Once the list has been defined, assess the value of each project for the organization under criteria such as cost, associated benefit, impact, and risk. From this information, prioritize, using a scale of 1 to 5, the various projects in your portfolio that the organization will implement.

STRETCH: Periodically add new projects to the list as they emerge from the ideas and problems stage of the process. Other elements of this activity may include creating a project task table for each new project that includes information such as the team, relationship with ideas, goals affected, schedule, and resource requirements.

REFLECTIONS

- Explain new product development.
- What are the main sources of funding for new product development?
- List the ways in which new product innovations can be protected.
- What are the key features of a market launch plan?
- What is the commercialization of new products?
- What are the main traits of an entrepreneur?

Table 9.1   Create Projects

Projects											
Title	**Cost**	**Benefit**	**Impact**	**Risk**	**Priority**	**Start**	**Due**	**Responsible**	**% Complete**	**Status**

Title: Title of the project

Cost: Cost of the project (e.g., $12k)

Benefit: Annual payback, revenue, or cost avoidance (e.g., 5k or a number from 1 to 5)

Impact: Impact of the project on goal attainment from 1 to 5

Risk: Level of risk associated with the project in achieving its impact from 1 to 5

Priority: Priority of the project from 1 to 5

Start: Start date of the project

Due: Due date of the project

Responsible: Person responsible for leading the project

% Complete: The percentage completeness of the project

Status: Status of the project (e.g., "not started," "in progress," "waiting," "completed")

197

10 Balancing Portfolios

Most established organizations have many innovation projects ongoing during their planning horizon in order to achieve their strategic objectives. A group of projects is called a portfolio. Applying innovation requires attention to the tools and techniques that can be used for managing and balancing a portfolio of projects. Projects will have various start dates, durations, due dates, assigned resources, and so on. A portfolio of projects is sometimes called a program or a plan. One of the key issues in managing a portfolio is achieving an appropriate balance or mix of projects for the organization. As investment budgets rise and fall and environmental conditions fluctuate, the fit of a specific project in the overall portfolio of projects will determine whether it should be implemented, shelved, or abandoned. This chapter deals with some of the important issues around managing portfolios of projects. The key difference between project and portfolio management is that portfolio management involves tradeoffs across many different projects, where the focus is on optimizing the achievement of organizational goals rather than the achievement of specific project goals.

LEARNING TARGETS

When you have completed this chapter you will be able to

- Explain the main ideas behind project portfolio management
- Describe four key strategies used in portfolio management
- Name a number of tools that can be used for portfolio management
- Explain the use of bubble diagrams as a means of evaluating portfolios

- Understand the difference between the portfolio dominant and project dominant approaches
- Discuss why organizations often have a mix of low-risk and high-risk projects

Portfolio Approaches

Portfolio management is about continuously choosing, managing, and adapting the mix of projects to match resource availability and contribute to organizational goals. All projects refer to a future state for a product, process, or service. Reaching that future state involves risk and can be unpredictable. Organizations are also dynamic, and therefore goals and priorities can change over time. For this reason managing individual projects can mean losing sight of the bigger picture. Portfolio management is an approach to managing the bigger picture—managing organizational goals and actions as a whole. There are four key approaches for portfolio management: maximizing the value of a portfolio, creating the right mix of projects, maximizing alignment with goals, and optimizing resources.

Maximizing the Value of a Portfolio

Maximizing the contribution of the portfolio involves placing a value on each of the projects in the portfolio. If all projects are assessed using a consistent approach, then it may be possible to compare the relative contributions of projects within the portfolio. Earlier we discussed a number of techniques for placing value on an individual project. Both qualitative and quantitative techniques can provide valuable information about the value of the individual projects. The use of solely quantitative techniques is inadvisable because it can produce a bias toward projects that have more directly assignable revenue streams. Using just financial techniques makes it difficult for the organization to justify projects that will build competencies and knowledge in platform technologies that will then be used to create future families of products. Although assessment based on financial techniques improves organizational efficiency in the short run, such a practice can severely limit the future flow of innovations along the action pathway. The alternative approach to optimizing the contribution of a portfolio is to include both qualitative and quantitative criteria in the assessment process. Together with the financial criteria, projects can be assessed on criteria such as strategic alignment, product advantage, market attractiveness, ability to leverage core competencies, technical feasibility, technological trajectory, and reward versus risk. Each project can be assessed using a score sheet similar to the one illustrated in Table 10.1. On this project's score sheet, teams award the project a score for each criterion. A project that achieves a high score on a criterion is generally seen to be more attractive than a project that receives a low score. In practice this type of selection model involves a group of experienced members of the organization who rate each project in the portfolio against consistent criteria. Each criterion can be given a weighting if the organization wants to prioritize one criterion more than others. Doing this can skew the portfolio mix in favor of the priority criterion relative to the others and in favor of certain project types. When the individual scores have been added up, the project receives a total score. The total scores for each project in the portfolio are then compared and an overall ranking of contribution is determined. Management can use this information to decide which new actions to add to the innovation portfolio and which ongoing actions should be changed or abandoned.

Creating the Right Mix of Projects

A disadvantage of maximizing the contribution of the portfolio is that it often selects "blue-chip" projects that have predictable but perhaps short-term and often moderate benefits. A complementary approach is to develop a mix of risky and rewarding projects. Risky projects can fail, but they may also provide significant benefits and rewards. The bubble chart is a simple tool for visualizing a mix of projects and providing decision support for managing a project portfolio (Figure 10.1).

Table 10.1   Score Sheet for an Individual Project

Criterion	Weight	Score	Weighted Score
Fit with mission	5	3	15
Impact on objectives	10	4	40
Impact on indicators	10	2	20
Fit with competencies	7	5	35
Fit with skills	5	5	25
Level of risk	10	4	40
Total score			**175**

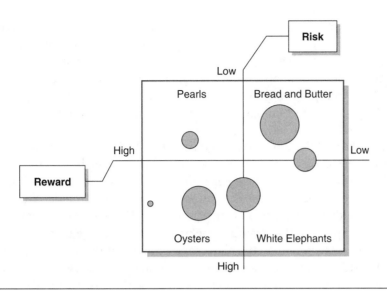

Figure 10.1   Bubble Diagram

This bubble diagram maps projects according to their impact on reward and risk. The size of the bubble in this instance represents the capital cost of the project. The position of the bubble indicates whether the project has the potential for high or low reward and a high or low risk. This presents the user with a visual decision support tool that allows projects to be ranked in the portfolio. There are three principal variables in a two-dimensional bubble diagram: the x-axis, the y-axis, and the size of the bubble. Another variable can be illustrated by changing the color of the bubble, for example, to indicate particular categories of projects. Some organizations also change the bubbles into ellipses that indicate, for example, probabilities. Each of the quadrants in a bubble diagram has a name that represents the relative value of projects in that quadrant:

Pearls: These are low-risk and high-reward projects. Clearly, these projects are highly desirable in any portfolio of projects.

Oysters: These are risky projects but with the potential of high reward. Sometimes the risk can be offset by the potential high reward.

Bread and butter: These are often small, simple projects such as continuous improvements. They have a high likelihood of success but make a low contribution to the overall benefits of the portfolio.

White elephants: These projects have low reward and high risk. These projects should be avoided but are often difficult to identify. Many projects start out as potential winners but become white elephants over time. It's estimated that about one-third of all projects, representing about 25% of spending, are white elephant projects. Organizations must be careful with respect to white elephants. First, it is never the intention to undertake this type of project at the outset, but because of the risks and challenges encountered, certain projects migrate toward this quadrant. Similarly, if reward is measured purely in financial terms, then certain projects may be deemed white elephants even though they may provide a technological platform for later innovations.

Another useful bubble diagram is shown in Figure 10.2. Here the organization has chosen to assess the spectrum across which its projects are focused. The projects are being examined relative to their project impact on both technology and market novelty. This mapping provides the organization with an overview of its future focus and also pinpoints gaps in its innovation plan. In this example, the organization can see that it has four major projects ongoing in the platform area and that the majority of their product support and derivative projects are small. This could indicate that the organization's existing product lines are reaching maturity as the organization is focusing most innovative investments on developing new platforms. If this is not the case, then managers must question why the majority of their innovation spending is so future oriented and what innovations they will provide to the market in the medium to short term. An additional variable can be added to the mapping by color coding the projects. This color code can be used to indicate variables such as phase of development, target market segment, or estimated time to market. The power of using a bubble diagram or the variation shown in Figure 10.2 is not only in the information it provides but also in the discussions that arise during its development. When members of an organization undertake a mapping similar to the one in Figure 10.2, they are able to visualize the types of innovations they will be delivering in the future. This provides them with an insight as to the likely evolution of the organization given the existing portfolio they are pursuing. If they are concerned with the way the organization is developing, then they may decide to change the portfolio mix by approving new projects of a certain type or eliminating existing projects from the portfolio. There are many different types of bubble diagrams that senior management can use to obtain alternative perspectives and assess their portfolio, including ease versus attractiveness, competence versus attractiveness, cost versus timing, strategy versus benefit, and cost versus benefit.

Maximizing Alignment With Goals

The third approach to portfolio management is optimizing alignment with goals. This strategy selects projects that are aligned with particular

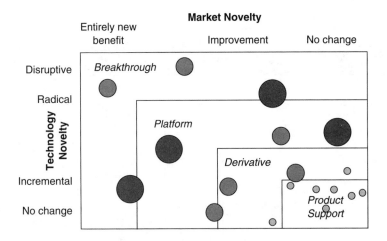

Figure 10.2   Scope Matrix

SOURCES: Adapted from Henderson (2007) and Wheelwright and Clark (1992).

goals, such as requirements, objectives, and indicators. Organizations undertake innovative projects to facilitate achievement of their stakeholder requirements and strategic objectives. Similarly, the competence of the organization to successfully undertake particular projects depends on the knowledge and expertise available for the project. Although alignment with these types of variables occurs during the approval assessment of individual projects, it is also necessary to examine the relationships at a portfolio level. A powerful but simple technique used to achieve this is the relationship or matrix diagram. Table 10.2 presents the relationships of an organization's project portfolio to its strategic objectives. The organization is then able to identify under specific variables where it has ample projects ongoing and where it may need to create new projects. The approval of projects should be determined not only by the attractiveness of the individual project but also by how it aligns with the needs and constraints of the organization's portfolio. The relationship diagram will be discussed in more detail in Part V.

Optimizing Resources

A fundamental element of effective management of the portfolio is management of resources. Optimizing resources is the process of balancing funding, worker hours, and skill requirements of the project portfolio with the resources available over a period of time. All projects consume resources, and every organization has only a finite amount of money available for

Table 10.2   Relationships: Objectives Versus Projects

Relationships	Projects							
Objectives	Install robotic welding	Redesign assembly line	Investigate ERP system	Develop workshop procedures	Restart sports and social activities	Implement innovation training	Implement e-auctions on selected items	⋮
Use low-disk strategy for capacity expansion		■	■	■				
Improve capacity analysis techniques			■					
Improve worker flexibility toward capacity changes	■							
Explore make-vs.-buy opportunities			■				■	
Collaborate on development of more accurate forecasts			■					
Explore manufacture-to-order processes		■	■					
Reduce order delivery times	■	■					■	
Improve dealer and supplier partnerships			■				■	
. . .								

innovation investment. In addition, the availability of funding often changes. For example, if revenues are particularly low, then the overall budget for innovation can be reduced suddenly. This can mean shelving particular low-ranked projects and placing emphasis on the higher-ranked ones. Management must assess the portfolio makeup relative to the available funds in order to determine the mix that will be most beneficial for the organization. Two other major constraints are used in determining resource optimization: time availability and skill availability.

TIME AVAILABILITY

The total number of hours a person has available to contribute to the project will influence the organization's ability to develop its actions. This is because a person often needs to balance the time spent on innovation activities with other day-to-day activities. Wheelwright and Clark (1992) highlight the case of work overload in their example of PreQuip Corporation. In this case, the examination of PreQuip's portfolio shows that management has overcommitted the available human resources by a factor of three. The implications are that there are too many projects ongoing in the portfolio relative to the available resources. The effect is that as projects are implemented, conflict occurs between teams as they struggle to access the necessary skills and resources. Such competition can lead to unnecessary conflicts within particular projects as teams engage senior managers to champion their cause and secure scarce resources. With the lack of necessary resources, the project focus switches from one of implementation and achievement of objectives to one of sustainability and survival. The overload results in suboptimal implementation of the overall portfolio as the organization struggles to implement all projects rather than a smaller number of properly resourced projects. In such a scenario, people experience significant stress as they try in vain to balance their day-to-day and project commitments. Interestingly, it is suggested that in order to maximize a person's value to the organization, the optimal number of projects for a person to be involved in is two (Wheelwright & Clark, 1992). The overloading of staff can lead to demotivation and reduce effectiveness. Table 10.3 highlights the resource loading of individual people engaged in the innovation process across the project portfolio and allows management to assess the available time each employee has to engage in existing and new projects.

SKILL AVAILABILITY

Even if an organization has spare human resource capacity in terms of time, it may still be constrained by the availability of necessary skills. The knowledge competence that can be drawn on to develop particular projects depends on the past project work of the organization and the experience of the available employees. Often projects are not approved because the organization lacks the foundation knowledge to make a project feasible. Similarly, if employees do not possess the necessary skills or are already fully committed to other projects, then the organization will be unable to undertake the project. Again, management must assess the attractiveness of the proposed project relative to the skills available. Where adequately skilled resources are unavailable, management must decide whether an ongoing project should be stopped or postponed in preference to the proposed one. A more detailed discussion of skills is provided in Chapter 11.

Table 10.3 Portfolio Resource Loading

Project Name	Start Date	Finish Date	Worker-Months								Project Total
			Ind 1	Ind 2	Ind 3	Ind 4	Ind 5	Ind 6	Ind 7	Ind X	
Install West-Tec	Jan. 07	Jul. 07	0	3	0	0	0	4	1	1	9
Relaunch Product X	Jan. 07	Nov. 07	3	0	2	0	5	1	1	3	15
Develop service Web interface	Jun. 07	Dec. 07	0	5	1	3	0	0	0	4	13
Implement Kanban	Jul. 07	Mar. 08	0	2	6	2	0	0	0	0	10
Develop new Product Z	Feb. 07	Sept. 08	5	5	2	5	1	0	3	0	21
Individual Total			**8**	**15**	**11**	**10**	**6**	**5**	**5**	**8**	

Portfolio Budgeting

When it comes to implementing a portfolio of projects, one of the key tasks is the preparation of budgets. Budgeting is an important mechanism for controlling and reviewing the progress of an innovation plan. The portfolio budget is also a proposed strategy for obtaining the necessary resources for implementing an innovation plan. The approval of project portfolio budgets reflects the policy and priorities of the organization because the allocated budget will determine the overall investment in terms of resources and commitment. There are two common budgeting methods: top-down budgeting and bottom-up budgeting.

TOP-DOWN BUDGETING

Top-down budgeting is a budgeting process in which one estimates the overall cost of the innovation plan and uses this estimate to allocate funding to individual projects and later to individual tasks within projects. As such, the budgeting process begins with an estimate of the amount of funding needed for the whole innovation plan (e.g., 4% of turnover). This estimate will be made on the basis of experience and judgment, together with any past data about similar projects. This primary budget is then allocated to projects using a variety of ranking techniques. Once allocated to projects, the budget can then be allocated to individual tasks. This process is illustrated in Figure 10.3. In this method, some projects will not receive funding because the budget may dry up.

A crucial factor for successfully implementing this method is the experience and judgment of those involved in producing the overall budget estimate. They must be able to take into account such aspects as likely time delays, inaccurate and changing cost estimations, and other resource considerations. One way of avoiding these problems is to request

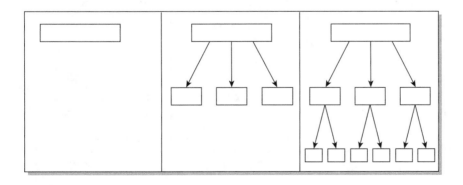

Figure 10.3   Top-Down Budgeting

a greater budget than is actually necessary. This freedom is often assumed when top-down budgeting is implemented, but many have found that this is not a safe assumption. Often, because of poor communication or external pressures, lower-level managers feel forced to accept the budgets allocated to them. This can lead to the budgeting process becoming a destructive game between managers in which one's gain is the other's loss. In practice, top-down budgeting is widely used and is well suited to traditional, hierarchically structured organizations. When carried out in the appropriate manner, it also has some important advantages, such as a high level of overall accuracy. It also provides a high level of stability in terms of the fraction of the total budget allocated to each project.

BOTTOM-UP BUDGETING

Bottom-up budgeting begins with identifying all the constituent tasks and projects that are involved in implementing an innovation plan and then summing up the resources and funding needed. This process of adding or aggregating the budgets of all tasks and projects is continued until a budget for the entire plan is arrived at. In the initial stages of a bottom-up budgeting process, the estimates are often arrived at through the basic tools for cost estimation; sometimes estimates can be in terms of worker hours or materials. When all the estimates are converted to cash, costs negotiations are often necessary between those responsible for each task and the project manager who is responsible for the overall budget. This is illustrated in Figure 10.4.

One of the main disadvantages of this method is that it leads managers to deliberately seek more funding than necessary. This happens because managers want to ensure that they will be able to accomplish the task to which they have been assigned and also because they surmise that all

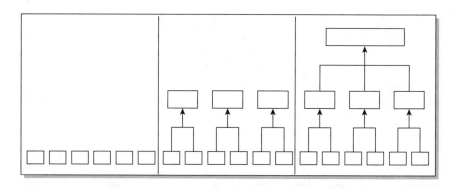

Figure 10.4   Bottom-Up Budgeting

requests for funding will be met only partially. This situation can lead to mistrust between the project manager and the other managers involved in the project. Another practical complication associated with bottom-up budgeting arises in drawing up a thorough list of all the fundamental tasks involved. It can be easy to overlook a task when working from the bottom up, and such an oversight can be the cause of serious errors in the overall project budget. The advantages of bottom-up budgeting lie in the accuracy of the budgets for individual tasks; this is useful from the point of view of controlling the project as long as all the tasks have been included. Other advantages stem from the level of involvement of personnel with responsibility for the various aspects of the project. In general, the estimates of the necessary funds and resources to complete each task will be more accurate if they come from those closely involved in performing the tasks. Employee involvement at this level can provide invaluable training and experience for those involved in the project and helps prepare them to take more responsible positions in future projects. In practice, however, the bottom-up approach to budgeting is rarely used, probably because of the role budgets play in the management process as a whole. Budgeting is both a mechanism for controlling projects and a source of power for top management. Therefore, top executives are unlikely to confer such power and control on inexperienced junior managers, especially because final accountability will still lie at the top.

Investment in Innovation

Projects consume resources: time, money, and people. Earlier we discussed the levels of investment that an organization makes in innovation. The level of investment varies across sectors, with the pharmaceutical industry spending approximately 18% of annual turnover, whereas the food industry spends an average of 1.5% (Trott, 2005). On average, an organization spends approximately 4% of turnover on various projects and initiatives linked with innovation. The expectation is that this investment will lead to improvements such as higher revenue, better quality, more intellectual property, and lower costs. Maintaining consistent investment in innovation is an important parameter of success in modern organizations. It is expected that the organization would at least recover its investment, and ideally make a significant improvement in revenues or service. Increased revenue and reduced costs can be collectively called benefit. One can argue that every project must achieve a reasonable cost–benefit ratio in the short or long term. The calculation of direct cost–benefit relationships associated with particular projects is often difficult to make and highly subjective.

The importance of continuous investment in innovation by organizations is highlighted by Trott (2005), citing a 1982 study (Wield,

1986) that examined the effects of ICI ceasing investment in product innovation. The study concluded that after cessation of investment, profits would decline slowly for approximately 15 years, but after this point they would drop off significantly. The study also examined the amount of time it would take for profits to recover from the initial collapse if investment in innovation were resumed at three times its original level. The study estimated that it would take another 25 years for the recovery to take place. Given the reduction in product life cycles and the increased complexity of modern innovations, it is probable that the effects would be even greater today, with the time to collapse being reduced and the time for recovery increased. Thus, continued investment in innovation by all organizations is essential to maintain competitiveness and sustainability.

Balancing the Portfolio

Each of the approaches discussed in this chapter can lead to a different selection of project portfolios. Maximizing contribution based solely on financial methods can lead to short-term, low-risk projects. A strategically aligned portfolio, on the other hand, may not yield any short-term benefits. There is also the question of the appropriateness of an organization's goals relative to the emerging environment. Strong alignment of the innovation portfolio with strategic objectives puts the organization at risk of not being aware of disruptive opportunities. Portfolio managers may need to be cognizant of the risks associated with screening and optimizing the innovation portfolio. Too rigid pursuit of the screening process can result in behavior such as abandoning the more radical of proposed innovations, focusing too heavily on a particular technology, or concentrating too much on existing markets. This type of behavior can be detrimental to the long-term competitiveness of the organization.

When the organization is deciding on its portfolio, it must balance the needs across the short, medium, and long term. The activity should be undertaken as a group exercise, and a number of the approaches listed in this chapter should be used to give as holistic and rich a picture of the portfolio as possible. The quality of the decisions about portfolio balance is directly related to the quality of the information managers have at their disposal. Projects and their place in the portfolio must be reviewed regularly to reflect shifts in the external and internal environments of the organization. One of the fundamental questions to be asked at this periodic review is whether the criteria used to rank the portfolio projects are still appropriate for the organization. The two systems of reviewing and assessing project suitability are the gates dominant approach and the portfolio dominant approach.

GATES DOMINANT

This approach focuses on reviewing each stage gate within the individual projects in the portfolio. At each stage gate a decision is made about future actions for the individual project. A regular portfolio view is sacrificed in favor of an in-depth review of individual projects. This approach is suitable where portfolios are static and is typically found in large, mature businesses. However, many modern organizations operate in fast-changing, turbulent environments that necessitate a portfolio-dominant approach.

PORTFOLIO DOMINANT

This favors a portfolio view over an in-depth review of individual projects. The portfolio and the organizational goals as a whole are reviewed regularly. The approach is particularly suitable in fast, dynamic organizations where projects are changing regularly and where the business environment is regarded as fluid (i.e., goals and projects are changing regularly). The adoption of a portfolio-dominant approach ensures that ongoing projects remain appropriate to the organization's emerging environment and changing goals. When an organization adopts a portfolio-dominant approach, it does not ignore stage gate reviews of the individual project. However, the organization does adopt a collective rather than an individual approach to its innovation effort.

Classification of Projects

There are a number of additional ways to classify projects and then use the classification information to judge the relative merits of projects across a portfolio. These classifications are also useful in providing insights into the type of innovations being carried out in organizations. Table 10.4 shows six different classification techniques. An organization may choose a number or all of these classification techniques for making decisions about a portfolio.

Some of these classifications are self-explanatory or have been discussed earlier. The "system affected" classification classifies the type of system affected by a project. For example, does the project make a major change to the information system, a business process, some technology, the organization, or a combination of all four? The "type of change" classification indicates whether a project makes a major change to capacity, compliance, maintenance, infrastructure, performance, or a combination of all five. The "stage of change" classification distinguishes between projects that are conceptual rather than, say, realizational. Finally, the "impact horizon"

Table 10.4   Various Classifications of Projects

System Affected	Scope of Change	Type of Change	Stage of Change	Impact Horizon	Scale of Change
Information	Incremental	Capacity	Concept	Containment	Core change
Process	Radical	Compliance	Development	Tactical	Next generation
Technology	Disruptive	Maintenance	Realization	Strategic	Upgrade
Organization		Infrastructure			Maintenance
		Performance			

classification indicates the more likely impact horizon for the project. Will implementation of the initiative achieve short-term benefits (i.e. containment) or medium-term benefits (i.e., tactical), or will the organization have to wait some time before benefits can be realized (i.e., strategic)? Many other classification techniques can be used for managing a portfolio of projects.

Summary

Organizations typically have a large number of projects. This is called a portfolio. The primary objective of portfolio management is to create a portfolio that will optimize the growth and success of the organization. There are four principal approaches to managing a portfolio: maximizing value, creating the right mix, maximizing alignment, and optimizing resources. These four approaches can be used together. Creating the right mix of projects can be aided by graphics such as the bubble chart and matrix diagram. There are two approaches to balancing a portfolio: the gates dominant approach and the portfolio dominant approach. There are many other ways of helping to decide which projects should be included in a portfolio and their relative rank. In dynamic organizations where goals are changing, portfolio management is often more important than managing individual projects to ensure continued goal attainment for the organization.

Activities

This activity requires you to create a bubble chart for your portfolio of projects using the data fields: risk, benefits, and cost. See example in

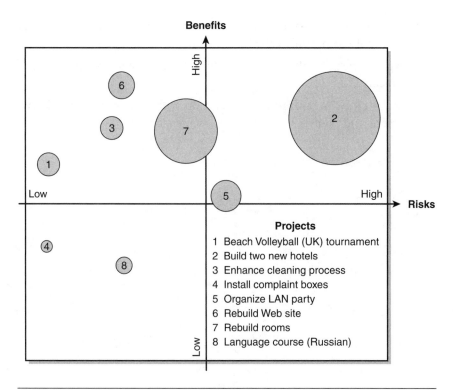

Figure 10.5 Create Bubble Diagram

Figure 10.5. Construct a bubble diagram for your entire portfolio of projects. The *y*-axis can represent the level of risk, the *x*-axis the reward or impact, and the bubble size the cost of undertaking the project. This may result in changes to certain activities that you have already completed. Revisit these activities and update if necessary. Your analysis may also lead to the generation of new ideas for innovative actions in areas where your portfolio is currently light.

STRETCH: Other elements of this activity may include creating a number of other types of bubble diagram for your project portfolio and incorporating them into your own portfolio management method.

REFLECTIONS

- Explain the main ideas behind project portfolio management.
- What are the four key overlapping approaches used in portfolio management?
- Name one tool that can be used for each of the four approaches.

- In a bubble diagram of risk versus reward, what are each of the four quadrants commonly called?
- Explain the portfolio dominant approach to balancing a portfolio of projects.
- Explain how organizations can end up with a large number of low-benefit and high-risk projects (i.e., white elephants).

Part IV

Empowering Innovation Teams

In the previous two parts of the book, we examined two of the key areas surrounding the innovation funnel: goals and actions. Goals set the objectives that drive the generation and implementation of actions by individuals in an organization. In this part we will look at the role these individuals play in the innovation process, particularly how they behave in a team. All individuals in an organization have a role to play in innovation. This can range from idea generation to participation in teams and can encompass the operative up to the leader of the entire organization. However, putting a team structure in place is just one objective; a much more important one is persuading individuals to behave as a team, to share a common purpose, to overcome obstacles, and to execute actions efficiently. There are three major components of ensuring the maximum contribution to such an environment: expertise, problem-solving skills, and motivation. Expertise comes from education and experience. In Part III we looked at a number of the more technical problem-solving skills. Social skills are another very important part of problem solving. The final component is motivation, and although individuals and teams are best motivated by intrinsic factors such as loving their work, extrinsic factors such as financial rewards may also play a role. Chapter 11 looks at some of the concepts of leadership in the creation of an environment that encourages engagement with innovation throughout the organization. Chapter 12 looks at building teams: how to create team structure and foster behavior that leads to more effective innovation. The final chapter in Part IV, Chapter 13, explores some ideas around motivating and rewarding team members within the innovation process. In particular, it looks at ways to connect an individual's goals with the overall goals of the organization.

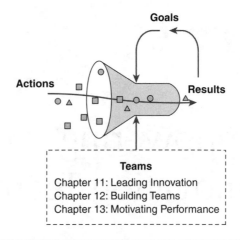

Figure IV   Empowering Innovation Teams

LEARNING TARGETS

When you have completed this part you will be able to

- Explain the importance of leadership in innovation
- Define the structure of innovation teams
- Understand how to build a team culture within the organization
- Describe some of the important attributes of virtual teams
- Understand communities of practice
- Explore various ways of increasing motivation
- Describe a simple performance appraisal system

Leading Innovation 11

L eadership is the ability to influence a group toward the achievement of well-defined, communicated, and accepted goals. Although leadership is a topic that has attracted much comment over the past century with regard to broader issues such as international conflict and nationhood, this chapter focuses on leadership as it influences innovation. The importance of leadership has been highlighted through the impact of people such as Steve Jobs at Apple, William McKnight at 3M, Jack Welsh at GE, and Herb Kelleher at Southwest Airlines. These people not only defined the goals for their organizations but also were the key drivers of many of the actions or projects undertaken. They also motivated other people to engage and act on their vision through the innovation process.

Leadership plays a critical role in stimulating innovation involving key people such as product champions and technological gatekeepers, together with the commitment and visible support of the organization's top managers. In the process of innovation, many actions originate from the bottom up; that is, ideas are generated by operatives, engineers, specialists, and users of a particular product, process, or service. What stimulates this activity is often the leadership skills that determine direction, motivation, championing, and support of the innovation process. Unfortunately, leadership skills are not easy to attain and often take many years of learning and experience. This chapter examines the leader's influence on innovation, how an innovation environment can be fostered across an organization, and what traits and styles contribute to making a successful innovation leader.

LEARNING TARGETS

When you have completed this chapter you will be able to

- Explain the leadership role in the innovation process
- Describe support roles played by individuals in the innovation process
- Understand a number of alternative leadership styles
- Explain how to develop an innovation culture
- Describe the difference between empowerment and responsibility
- Discuss some of the key issues of conflict resolution
- List a number of core competencies necessary for successful leadership

Transformational Leadership

Leadership can be viewed as either transactional or transformational. Transactional leadership is typically used in established organizations (Table 11.1). The style emphasizes attention to short-term goals, the need for formal structures, solving problems as they occur, and essentially maintaining the status quo. Transformational leadership, on the other hand, draws attention to long-term visions, encourages empowerment and trust, and focuses continuously on the need to change existing products, processes, and services and, over time, the innovation process itself.

Table 11.1　Leadership Styles

Transactional Leadership	Transformational Leadership
Clarify goals and objectives to obtain immediate results	Establish long-term vision
Create structure and processes for control	Create a climate of trust
Solve problems	Empower people to control themselves, manage problem solving
Maintain and improve the current situation	Change the current situation
Plan, organize, and control	Coach and develop people
Guard and defend the culture	Challenge and change the culture

SOURCE: Adapted from Schein (2004).

Transactional leaders often are regarded as conservative bureaucrats who abide by the rules of the organization. A transformational leader, on the other hand, is viewed as a maverick who continuously challenges established authority, attempts to seize every opportunity, questions every rule, and motivates and controls people through personal loyalty. When discussing the relationship between leadership and organizational transformation, it is important to stress that although a charismatic leader has strengths in envisioning, energizing, and enabling innovation to occur, it is often not enough for sustained innovation (Nadler & Tushman, 2004). An organization may view its reliance on transformational leaders as a weakness leading it to pursue unrealistic goals. An environment of conformity may also exist, and there will be a reluctance to challenge the direction of the leader. In addition, lower management layers may be disenfranchised by the practices of the transformational leader. Although these problems can thwart the innovative efforts of an organization, effective leaders will build competent teams and structures around themselves, develop successful reward and control systems, and constantly strive to clarify the behavior required of their followers (Nadler & Tushman, 2004). Consequently, the effective leader combines the best elements of transactional and transformational styles of leadership by building structure and systems to reinforce his or her charismatic abilities. The management of the innovation process requires elements of both styles of leadership at different stages of the life of a product, process, or service. It needs the transformational, charismatic leader in order to generate the desire and environment for innovation, but it also needs more structured transactional leadership to address ongoing operational issues within the innovation process.

General Leadership Traits

Leadership is important for all activities in an organization, not just innovation. Although significant research has been undertaken over the years to determine the characteristics or traits that create good leaders, no ultimate formula has been defined. That said, certain traits have been demonstrated to be advantageous to a leader's success. The following traits have been identified as being important (Robbins, 1998): ambition and high self-motivation, the desire to lead, honesty and integrity, self-confidence and intelligence, job-relevant knowledge, and the ability to delegate and listen.

AMBITION AND HIGH SELF-MOTIVATION

A person's drive to succeed can challenge existing norms and identify opportunities that others have failed to see. These people are high achievers and can significantly advance the performance of the organization.

THE DESIRE TO LEAD

In this case, the person must relish the responsibility and challenge of the role of leader. Ambition, self-belief, and desire for a challenge are attributes that play a very large part in a person wishing to don the mantle of leader.

HONESTY AND INTEGRITY

To be a successful leader, one must be able to communicate and motivate others and therefore be capable of gaining trust and confidence. If the leader possesses honesty and integrity, he or she will be more likely to attract those who will follow his or her direction.

SELF-CONFIDENCE AND INTELLIGENCE

Having the desire to lead denotes a high level of self-confidence; the person usually relishes any challenges that may arise. Experienced and intelligent people are often obvious choices as leaders because they inspire the belief that their knowledge will help avoid potential pitfalls and facilitate success.

JOB-RELEVANT KNOWLEDGE

Experience and a track record in the relevant field are other criteria that make a good leader; these traits can inspire others in the organization to believe in the leader's success. Often when the organization is faced with technological challenges, relevant knowledge on the part of the leader is crucial to success.

ABILITY TO DELEGATE AND LISTEN

No leader can go it alone, and good leaders surround themselves with excellent people who will support them in achieving their objectives. Thus the art of delegation and also the ability to listen to alternative opinions and feedback must be learned. In this way, the leader validates his or her vision of the future and can alleviate concerns among colleagues before they develop into resistance.

Although this list is by no means exhaustive, these traits contribute to individuals becoming better leaders. Leadership rests on one fundamental fact: Leaders have followers. Each of the traits defined earlier can attract followers, who may adopt a leader for a number of reasons:

- They have confidence in the leader to deliver what they cannot achieve alone.
- They empathize and connect with the leader's values and beliefs.
- They are attracted and motivated by being associated with the leader.
- They can easily attribute authoritative power to the leader.

Although in certain organizational scenarios the leader will need to adopt a top-down approach to decision making, more often leaders adopt a consensus approach of engaging others in the decision-making process. In doing so, they can distill the collective wisdom and experience of the group and use this information to make more informed decisions. Similarly, by engaging others in the decision-making process they increase the chance of support across the organization as a whole.

Innovation Leadership Traits

Leadership is critical to innovation success because it counteracts organizational decline by creating a sense of urgency in employees to take action (Kotter, 1996). An organization benefits from leadership as people are motivated to channel ideas and solutions into fulfilling organizational goals. In this way, the organization moves toward a position of greater competitive advantage as concepts are nurtured and supported throughout the innovation process. The role of leadership with regard to innovation consists of the following responsibilities (Lampikoski & Emden, 1996): vision and goal setting, developing core competencies, motivating people, and nurturing ideas.

VISION AND GOAL SETTING

The first influence leadership has on innovation is to define the direction in which the organization will proceed into the future. Leaders create a vision and set future goals for the organization. This vision, inspired by the leader, will guide the creative efforts of all people. The development of these visions and goals must be coupled with an excellent communication structure that allows all involved to understand and embrace their own goals and how they relate to the vision. Effective leadership channels these goals down through all layers of the organization, ensuring that all suborganizations coordinate their efforts to facilitate goal achievement. Good leaders achieve buy-in to ensure that their vision is shared by all. Leaders demonstrate their own commitment and support for goals and intuitively know the actions needed to move toward their vision. Although leadership can be a powerful enabler of innovation, it is important to highlight that in certain circumstances it can also hinder it. This can happen if a leader adopts a vision that is myopic or out of alignment with the emerging environment. This can also happen when the organization's leaders promote a philosophy of "not invented here" that rejects potential innovations because they originated from sources outside the organization; this leadership practice can significantly reduce the potential opportunities available for exploitation (Tidd et al., 2005). An important aspect of good leadership is knowing when to abandon current organizational goals and, even more important, to develop new, more appropriate ones.

DEVELOPING CORE COMPETENCIES

The second sphere of influence a leader has on the innovation process is ensuring the ongoing development of core capabilities that provide the foundation for innovation in the future. The organization's leaders must ensure not only that the appropriate vision for the future is identified but also that the organization has the ability and competence to undertake related innovation actions that realize the vision. The role of a leader is to find the correct balance between the short-term and long-term skills needed to support innovation. This can result in investment in innovation by senior managers in order to provide the technological and knowledge platforms that allow future innovative actions to be pursued. Leaders must also support training and development, especially in areas where skilled people need retraining.

MOTIVATING PEOPLE

This is a third area where organizations can gain significant competitive advantage through effective leadership: their leaders' ability to motivate employees to move out of their comfort zone and become actively engaged in the innovation process. This not only increases the skill base and resources available to advance innovative actions but also increases the range of contributors to the idea generation phase, leading to more diverse innovative concepts being considered by the organization. Many leaders motivate people by adopting an empowerment strategy in which they not only delegate responsibility for tasks to the lower layers but also provide the necessary training to enable them to fulfill their new responsibilities. Motivated people demonstrate greater commitment and can significantly reduce the resistance encountered by a proposed innovation as buy-in is achieved more easily. In this way the organization may go on to generate a greater number of innovative concepts, and over time these motivated people may be empowered to become champions of innovation in their own areas and will even begin to motivate others.

NURTURING IDEAS

A fourth role of effective leadership of the innovation process relates to supporting and nurturing innovative initiatives across the organization. In the past, innovation was the domain of the R&D or design department. However, with the popularity of empowerment as a management philosophy, innovative concepts can now originate from any person or location in the organization. Leaders need to develop an environment that will nurture and support the development of innovations, regardless of whether they originate within or outside the organization. Tidd et al. (2005) identify aspects of leadership such as shared vision of the future, extensive communication, the desire to innovate, and the achievement of high involvement in the innovation process as key components of an innovative organization. Nurturing innovation is linked to ensuring that people are motivated to engage with the innovation process. This can be achieved through communication, education, and training or by linking innovative behavior with financial rewards and appraisal systems. Leaders must ensure that organizational systems and structures support the innovation process. This can result in processes that structure the flow of innovative actions through the various stage gates or that correlate training and development with the skills needed for the innovation process. Managers can also demonstrate leadership in supporting innovation by ensuring that actions have the resources needed for effective implementation. This may often involve having to make tradeoffs at the various stage gates of the innovation process in order to decide the most appropriate action to progress. Leaders can also support the development of innovations through their roles as sponsors and champions within the innovation process. Ultimately they must ensure that the process works, is transparent and well understood by all, and is efficiently managed so that innovations can happen. An organization benefits from leadership as employees are motivated to channel their ideas and solutions into fulfilling the organizational goals. In this way, the organization moves toward a position of enhanced innovative capability as concepts are nurtured and supported throughout the innovation process.

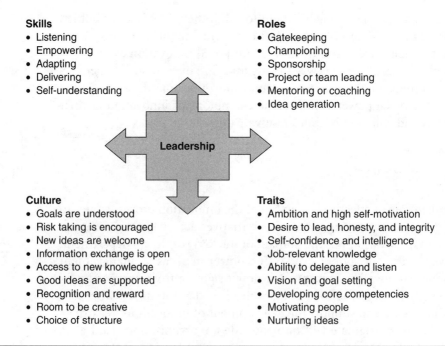

Skills
- Listening
- Empowering
- Adapting
- Delivering
- Self-understanding

Roles
- Gatekeeping
- Championing
- Sponsorship
- Project or team leading
- Mentoring or coaching
- Idea generation

Culture
- Goals are understood
- Risk taking is encouraged
- New ideas are welcome
- Information exchange is open
- Access to new knowledge
- Good ideas are supported
- Recognition and reward
- Room to be creative
- Choice of structure

Traits
- Ambition and high self-motivation
- Desire to lead, honesty, and integrity
- Self-confidence and intelligence
- Job-relevant knowledge
- Ability to delegate and listen
- Vision and goal setting
- Developing core competencies
- Motivating people
- Nurturing ideas

Figure 11.1   Leadership Attributes

Leadership Roles in Innovation

Leadership in the innovation process is not the sole domain of the senior management team. Instead, leaders must be present at all levels of the organization if innovation is to become systemic. A number of critical roles where leadership is necessary to support the innovation process include gatekeeping, sponsorship, championing, project or team leading, mentoring or coaching, and idea generation. See Figure 11.1 for these and other leadership attributes.

GATEKEEPING

Leaders must ensure consistent decision making at the various stage gates of the action pathway. They must ensure that the innovative actions that are allowed to flow through the various stage gates are aligned with the goals and constraints of the process. Ultimately, leaders must determine the appropriate mix and focus of actions that will result in future innovative success. Similarly, as gatekeepers, leaders must act as knowledge conduits to pass on related developments of the external environment to those in the organization. Leaders must also strive to improve the way in which innovative ideas move through the various stage gates; by doing this, they help improve the effectiveness of the organization in developing innovative processes.

SPONSORSHIP

People at different levels of the organization can act as sponsors for specific innovative actions or, more generally, promote the need for ongoing innovation. Because many organizations can suffer from inertia, promoting innovations and the subsequent changes can be difficult and can encounter resistance. In order to overcome this, sponsors can use their power base within the organization (either reputation or position power) to advance the action by helping secure the necessary resources and approvals. This can sometimes lead to the organization questioning the appropriateness of its current goals and competencies relative to the future environment. The sponsor can give legitimacy to a project by supporting and protecting it from unnecessary bureaucracy and politics. Often the sponsor can also adopt the role of mentor to the innovation team.

CHAMPIONING

Leaders are needed to champion potential innovations into existence. Innovations disturb and disrupt the status quo and involve risk. Therefore, people are needed who believe in the innovation enough to champion its cause across the organization in order to sell the concept to others. Champions demonstrate commitment and leadership by exposing themselves to failure if the project fails. They engage others to support innovation and drive the action forward from concept to innovation with their motivation and commitment.

TEAM LEADING

Innovative actions are implemented in projects, and these projects must be managed to ensure optimum success. To achieve the project objectives, the project manager or leader must coordinate the team and available resources. He or she must communicate goals to be achieved, motivate team members, overcome barriers to progress, and secure adequate resources to allow successful implementation. The project leader must champion the innovations across the organization and ensure that the project has the necessary sponsorship and support. In some instances, the project leader and original champion of the concept may not be the same person; this may occur if the original champion lacks the management skills necessary for the project, resulting in a different project leader being appointed, or similarly if the champion is the sponsor of the project.

Either way, the project leader must demonstrate leadership to ensure that the team operates effectively and the project is implemented successfully.

MENTORING AND COACHING

Leaders are needed to support and nurture innovative actions to fruition. These key people in the organization provide teams with knowledge, insight, and experience that assist in bringing success to innovative projects. They liaise with the project leader and highlight the pitfalls to be avoided. Access to this expert knowledge facilitates the development of the innovative action and increases organizational support for the project.

IDEA GENERATION

Creativity is a fundamental element of the innovation process. Often, the greater the number of innovative concepts being submitted to the "fuzzy front end" of the innovation process, the better the chances of success at the end. An innovative organization needs all people to exhibit leadership in developing creative ideas and innovative solutions that can then be inputted to the innovation process. People must move outside their comfort zone and engage in this process through idea generation and participation in project teams to make their innovative ideas a reality. This is often the result of an empowerment culture driven by senior leaders.

Leadership Skills

A complementary and overlapping perspective on leadership is to look at the skills that can be usefully developed. A comprehensive set of skills for effective leadership can be grouped around the following areas (Jones, 1996): listening, empowering, adapting, delivering, and self-understanding.

LISTENING

This is a critical aspect of leadership because subordinates respond much better if they feel that someone is listening to them and that they can have input into a process. Listening is linked to individual motivation. If people feel free to suggest ideas without fear of being dismissed or ignored, this provides the organization with a huge network of new perspectives that may lead to innovation. Specific skills include being able to

- Listen carefully to others
- Encourage ideas from the team
- Motivate and encourage others
- Develop a good communication network

EMPOWERING

This allows leaders to renounce certain areas of power, giving them the space to lead more effectively. In innovative organizations, empowerment and the ability to lead empowerment initiatives are essential. Specific skills for leaders include the following:

- Giving people responsibility for tasks and projects
- Demonstrating trust
- Providing training to enable people to work effectively
- Providing support for people where needed

ADAPTING

This refers to challenging the accepted norms of the organization and pushing the boundaries of accepted situations. In this period of continuous change, leaders must adapt to circumstances as they occur. Specific skills include being able to

- Challenge the rules and conventions of the organization
- Anticipate and adapt to changing conditions
- Help others to manage change
- Manage stress well

DELIVERING

This refers to delivery of the organization's vision and to demonstrate complete commitment to its attainment. It is through a leader's organization and motivation of teams that actual results are delivered. Specific skills for a leader include being able to

- Have a clear vision for the team
- Communicate the vision and ideas clearly
- Demonstrate a high level of commitment to work
- Focus on achieving results

SELF-UNDERSTANDING

This is crucial; leaders need to recognize their own skills and focus on acquiring other effective leadership skills. Leaders must demonstrate "soft" skills, such as coaching, as well as "hard" skills, such as competency with technology. They must demonstrate a positive attitude toward themselves and their work. Specific skills include being able to

- Have a clear perception of personal strengths and weaknesses
- Spend time keeping up-to-date and developing new skills
- Manage time well
- Have a positive attitude

EXAMPLE: William Hewlett and David Packard created an open, decentralized Hewlett-Packard corporation, priding itself on innovation and creativity. This open culture encouraged new ideas and allowed the company to become a world leader across a range of products. In 1990, when the founders retired, the company was becoming more centralized and bureaucratic, perhaps the inevitable result of its global success. Innovators found that they needed to get approval for new ideas from many layers of management and received much less moral support to continue working on the new ideas. The result was a marked slowdown in new product introductions and falling profits. Hewlett and Packard returned from retirement and broke up the bureaucratic mess. They reintroduced a positive environment for ideas and risk taking. The result was that new product introductions increased, as did profits.

Leadership and Culture

The culture of innovation was introduced earlier. We now extend this discussion in the context of leading innovation. Culture can be defined as the pattern of shared basic assumptions that an organization uses in order to deal with internal and external shifts (Schein, 2004). Because culture reflects the fundamental beliefs of the collective, the organization can harness this potential as a powerful enabler. Organizations that develop a culture that is aligned with the requirements of innovation can nurture innovative actions and increase their innovative capability. However, organizations that have a culture that has a poor alignment can stifle and impede innovations, reducing the likelihood of success; these organizations will need to work on changing their culture to a more supportive one.

An organization's capacity to innovate is very much influenced by leaders because they mold the environment in which innovation happens

(Goffin & Mitchell, 2005). New ideas will succeed only if the organizational culture allows them to grow and prosper. Thus, one of the key duties of a leader is to establish an organizational environment that nurtures innovation. There are many factors necessary for developing an effective innovation culture in any organization, including the following:

- Goals are understood by all.
- Risk taking is encouraged and accepted.
- New ideas are welcomed.
- Information exchange is open and shared.
- Access to new knowledge is extensive and uncontrolled.
- Good ideas are supported.
- Innovations are recognized and rewarded.
- People are given room to be creative.
- There is a choice of structure.

UNDERSTANDING OF GOALS

As discussed earlier, one of the fundamental roles of leaders is the defining of goals. In order to develop an environment where all people engage in the innovation process, leaders must communicate defined goals to all and ensure that their importance and purpose for the organization's future are understood. Kotter (1990) states that leadership is a process, whose purpose is to help direct and mobilize people and their ideas. Therefore, the leadership style must provide direction for the organization and allow people to feel that they have input into the development of strategies for the future. This communication is important for organizational innovation because it achieves buy-in and allows people to align their idea generation with the requirements of the organization or challenge organizational goals they feel are inappropriate.

RISK TAKING

Taking risks is a necessary part of generating innovations. Although many ideas will fail, a few will succeed to become innovations. It is the success of these few that must justify the entire innovative effort. The risk–benefit relationship must be an integral part of the skill set of every innovation leader. Two strategies allow leaders to lower the impact of risk: diversification and hedging. Diversification encourages many new ideas to be developed simultaneously. If there are many different ideas, the chances are far greater that one of these will be a success. Hedging allows new ideas to be developed to a point where decisions must be made about additional resources. This is equivalent to a gambler placing low bets at the beginning of a game of cards, then deciding at appropriate milestones whether to fold or bet even more. Leaders must develop a culture where risk taking is encouraged and where failure (though not actively sought) is accepted as a learning opportunity for the future.

WELCOMING NEW IDEAS

This is consistent with the transformational management approach, in which any ideas that will lead to progress are welcomed. This is not to be confused with pursuing all ideas; poor ideas or ideas that have an unacceptable level of risk may need to be terminated. As discussed earlier, appropriate termination of ideas is also part of the innovation process. New ideas must be welcomed if they have strategic fit and if the resources are available to bring them to successful completion. Idea generation and termination must be seen as the rule rather than the exception. At meetings, time must be allocated to allow new ideas to be suggested and openly discussed. If a decision is made to terminate an idea, the reason should be communicated to the relevant person in order to avoid demotivation and to encourage the generation of more suitable ideas in the future.

INFORMATION EXCHANGE

This is the lifeblood of innovation. Who is responsible for what goal? Who is working on a particular idea? Who has skills in a particular area? What is the emerging state of the art in the area? What does the market need? Many ideas depend on the thoughts, skills, encouragement, and knowledge of a number of people. Leaders can encourage open information exchange through regular meetings, workshops, and online collaboration that can enhance creativity and problem solving.

ACCESS TO NEW KNOWLEDGE

The Internet has created an explosion of new knowledge sources. In the past, organizations relied on expensive libraries, subscriptions to magazines, and trips to conferences and trade shows. New knowledge is the raw material for creative thought. Many techniques are used for capturing and sharing new knowledge, including building communities of practice with other like-minded people, using the Internet, and attending training seminars and university courses. All of these contribute to developing innovative and cogitative skills in the workforce.

SUPPORT FOR IDEAS

Ideas are like young seedlings: If they do not receive support and encouragement, they can wither and die. Innovation leaders need to learn how to support ideas and protect the often delicate sensitivities of the people who create them. Innovation leaders play a number of important roles in this process. First, they receive the idea sensitively and encourage its development during its initial growth stage. As the idea grows, so will its demand for more resources and its impact on other managers. Innovation leaders often need to defend and promote ideas with their peers and champion the idea to more senior management teams and boards. Through such support, a concept develops into an action that benefits the organization.

RECOGNITION AND REWARD

Innovation and creativity demand appropriate rewards and recognition for the person or team that creates and develops the innovations. This is necessary to encourage them to continue to engage in the innovation process. Creative energy is easily dissipated but can be replenished by using reward and recognition. Rewards can be either intrinsic (e.g., self-actualization) or extrinsic (e.g., financial incentives). Both are interlinked. A key issue is to not allow extrinsic rewards to overshadow the intrinsic rewards, which are often more meaningful and satisfying. By linking reward and recognition with the innovation process, leaders communicate its importance to all.

SLACK

Both creativity and the innovation that may result need significant time and space to occur. The majority of organizations struggle to find the appropriate balance between the day-to-day pressures of operating a business and the future-focused pressures of the need to innovate. An organization often finds itself in a situation where the majority of focus is on short-term operational pressures, to the detriment of future innovations. Leaders must create slack in day-to-day routines to allow experimentation, investigation, and creativity. Time out from standard duties is also needed if people are to work in teams to move innovative projects to successful conclusions. Without strong leadership in this regard, responsibilities will increase, but the time and resources needed to follow innovations through will decrease. Creating an environment with a certain degree of flexibility and spare capacity allows people to pursue ideas and problems that may in time lead to significant innovation.

CHOICE OF STRUCTURE

This will also have an impact on the culture and innovative ability that an organization develops. A flat structure in which people can communicate easily and operate as a group is the most advantageous style. Although it is sometimes beneficial to invest responsibility with just one person, such an individualistic approach negates the advantages of teamwork (Katzenback & Smith, 1993) and limits a person's cognitive space. A more consensus-based approach to the management and direction of the organization can be obtained by operating as a team.

EXAMPLE: The 3M Company has been in existence for more than a century and has produced numerous innovations, including masking tape, Post-it Notes, and Scotchgard. Innovation has been at the heart of the organization since its inception, and consequently its senior management has succeeded in developing an environment within 3M that nurtures and supports innovative endeavor. The 3M culture has been developed through constancy of purpose that has affected the policies, processes, and practices pursued by the company. These routines include the following:

- A conscious effort to recruit creative, independent people with specialist knowledge to support their divisions.

- A company mantra of "better to ask forgiveness than permission" that encourages people to pursue ideas they believe in even if not supported by their immediate boss.

- A 15% rule, which allows people to devote up to 15% of their work time to innovative initiatives they think will benefit the organization. This slack gives people the opportunity to experiment with concepts and make discoveries that may have commercial potential.

- A goal that 30% of the company's annual turnover must be generated from products that are less than 4 years old. This ensures that all employees understand the importance of continuously developing new products rather than resting on past successes.

- Active training and development of all people to maximize their contribution to organizational sustainability.

These and other routines have been developed and implemented by 3M's leaders to ensure that innovation remains at the heart of the company and that their process will continue to produce innovations that are embraced by the market.

Conflict Management

Leadership is needed to resolve or avoid conflict that can occur as a consequence of introducing an innovation that threatens the status quo. Diversity is an important attribute of an innovative team. It fosters creative friction between people that spurs new ideas and goal attainment. However, this friction can also lead to conflict, and the purpose of a leader is to make this conflict positive rather than destructive. Leaders can manage this conflict by creating the right climate, facilitating discussions, and achieving closure (Luecke, 2003).

CREATING THE RIGHT CLIMATE

This involves creating an environment where obstacles and problems can be discussed as they occur. Each member of the team needs to be encouraged to expose relevant issues without fear of recrimination; these could include interpersonal team issues or fear of exploitation.

FACILITATING DISCUSSIONS

When issues are raised, it is important that the team leader facilitate the free expression of views by keeping the discussion on an impersonal basis. The leader must carefully manage this discussion by constantly referring back to the team's objectives and goals in order to prevent tangential issues from impeding a resolution.

ACHIEVING CLOSURE

Closure can occur when all parties involved in the conflict come to an agreeable solution. However, this may not be achievable within the constraints of one meeting. Where discussion is leading nowhere, leaders have the option to park the issue and revisit it later when people have had time to consider alternative solutions.

EXAMPLE: HarPer Sculpting Corp. provides a diverse range of health and fitness options to its customers. The organization has created a strategic plan and has allocated responsibility and team membership to many of the people in the innovation team. The matrix shown in Table 11.2 shows the relationship between the upper management personnel and the objectives of the company. Here we can see the main responsibilities of each manager and an assessment of how they perform their duties. Although some are

Table 11.2   Individuals Versus Objectives

Relationships		Objectives								
Job Title	**Name**	Attain annual growth	Conduct in-depth analysis of finances	Improve analysis of new markets and locations	Pioneering regimes in the health and fitness sector	Measure customer satisfaction	Improve workforce through training and education	Provide safe e-commerce for customers	Achieve an empowered and happy workforce	Conduct improvement meets with increased employee involvement
Operation manager	Ivan Fallon		●			●			●	●
CEO	John O'Malley	◺	◺							
R&D	Lolita LeVeaux			●	●					
Advertising and e-commerce	Martin Casey		●							
HR and safety manager	Mary O'Houlahan						●	●	●	
Site manager (Cork)	Michael Divily					●	●		●	●
Site manager (Dublin)	Patch Ludwig					●	●		●	●
Financial manager	Paul McGoldrick	●	●	●						
IT manager and customer service	Peter McNamara					●		●		
Site manager (Galway)	Tom Griffin					●	●		●	●

Legend
Involvement:
  Strong ● (Strong)
  Medium ○ (Medium)
  Weak ◺ (Weak)

deemed weak, ongoing training and education in these areas should ultimately improve the overall performance.

Summary

Innovation leaders are driven by vision, ambition, and energy. They are transformational risk takers and constantly change the status quo. They must also exhibit traits of more transactional leaders who nurture the more practical and routine issues of innovation management. Innovation leaders foster a culture of innovation that encourages appropriate risk taking, welcomes new ideas, shares information openly, and recognizes the contributions of individuals. Conflict provides a creative friction between individuals but must be managed carefully so that it remains positive. Innovation leaders are equipped with the skills to manage conflict; these include facilitation and task management. In the next chapter we will examine the followers of innovative leaders, namely the people who make up the teams that implement innovation. We will investigate the structures by which they, together with innovative leaders, form effective teams to develop innovative actions throughout the process. The importance of empowering individuals to pursue innovation is also discussed.

Activities

This activity requires you to assign responsibility to the people you allocated earlier for each of the goals and actions in your innovation plan and check for consistency and appropriateness of your assignment. Be careful not to allocate people at random; consider why each person needs to be allocated to a particular goal or action. Your allocation must be credible and realistic. Check the various people for appropriateness in terms of skills and experience and also their availability in terms of workload before assigning them responsibility. To complete this activity you will need to use a simple matrix diagram. Copy Table 11.3 into a spreadsheet. Create three versions of the matrix: "Objectives Versus Individuals," "Indicators Versus Individuals," and "Projects Versus Individuals." Shade the intersecting cell where a person is responsible for a particular objective or project. Use a different color to indicate where a person is not the leader but is part of the team. Save your spreadsheet file. You will need it later when constructing your innovation plan.

STRETCH: Other elements of this activity may include creating a resource loading model for the various people you assigned and the various activities they are assigned to.

Table 11.3   Assign Responsibility to Individuals

Relationships		
		<List 2>
<List 1>		

List 1: Individuals

List 2: Indicators, objectives, etc.

To paste a list into the <List 2> cells, first copy the data you want to paste,

then select the first cell in the <List 2> cells above,

then select Paste Special.

In the dialog box, click on the Transpose box and then OK.

Now select all the cells in <List 2> and select Cells from the Format menu.

Select the Format tab and change orientation to 90 degrees. Click OK.

REFLECTIONS

- Why is transformational leadership desirable for effective innovation?
- Describe the different traits leaders should possess in the innovation process.
- List six key factors for fostering an innovation culture in any organization.
- Describe two different leadership styles.
- List the interpersonal skills for leaders.
- Explain in your own words why conflict might be a positive influence for innovation.
- Identify the various leadership roles in the innovation process.

12 Building Teams

In large organizations there will be many teams engaged in applying innovation. The building of effective teams ensures that those involved in developing innovative actions or managing other aspects of the innovation process have the necessary skills, knowledge, and autonomy to achieve their objectives. The presence of these teams facilitates success in achieving innovation goals. The traditional approach to organizational design is to divide a large organization into functions or departments. Each department can be an innovation team. In addition, innovation teams are drawn from one or more of these departments in order to define new goals, create ideas, and implement projects. Other types of teams are also involved in the innovation process. Management teams can be innovation teams that determine the strategic direction for the organization, coordinate the activities of various functions, and ensure cohesion between the ongoing actions. This chapter looks at ways to define innovation teams in a large organization. We begin by looking at the traditional structure of an organization. We then look at other approaches to structure, such as cross-functional organizations and virtual teams. Most organizations now operate a type of matrix organizational structure in which individuals report on a permanent basis to functional departments but also report on a temporary basis to project managers. Functional departments can be examples of permanent teams. Once formed and structured, these teams can last for many years. Projects, on the other hand, are examples of nonpermanent teams and last only for the duration of the project. The innovation process requires both types of teams to operate effectively to achieve objectives. This chapter looks at teams and how they are used to stimulate, plan, execute, and control an innovation plan.

When you have completed this chapter you will be able to

- Explain the different types of team structure in organizations
- Define various types of permanent and nonpermanent teams
- List successful traits of highly effective teams
- Describe a number of ways to enhance creativity in teams
- Explain the concept of virtual teams
- Understand the concept of communities of practice

Organizational Structure

The human resources in an organization can be its greatest competitive advantage and are essential to generating and developing innovative actions. Organizational structure is a system for organizing roles and responsibilities into more manageable parts. The most common form of structure is to divide an organization into departments, as illustrated in Figure 12.1. Departments allow groups of people to specialize and manage one particular aspect of an organization's activity, usually a function such as engineering, marketing, design, or operations. Specialization allows organizations to develop experience and share knowledge in a manageable way. People are typically assigned to one department, which also contains one overall manager. This type of structure leads to a functional hierarchy in which each department understands its role in meeting the overall objectives of the organization. The structuring of departments and subdepartments varies widely between organizations and even within the same organization.

In addition to individual departments or subgroups within departments, other types of innovation teams are also present in organizations; management teams are the most important of these. Other types of teams include project teams, cross-functional teams, and strategic partners across organizational networks.

Departments Responsible for Change

Many organizations have departments, which specialize in changes to information systems, process technology, and product design, and it is

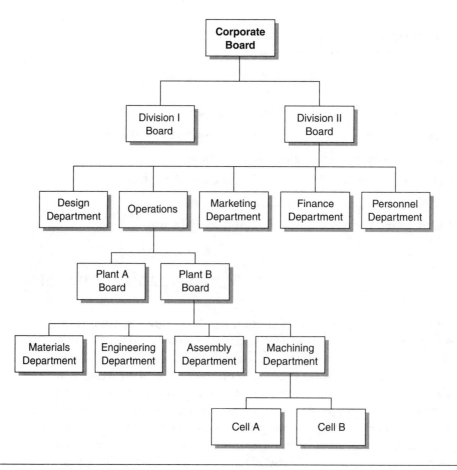

Figure 12.1   Departmentalization

these departments that typically take responsibility for a significant proportion of innovation and change. However, as we saw earlier, innovation is often a multidisciplinary activity involving people from a number of departments. Nevertheless, it is worth considering these three special departments in a little more detail. The departments of information systems and process engineering share a special relationship because information systems are applied to processes, and process engineering uses information systems. This is illustrated in Figure 12.2. Both departments are seen as overlapping circles, with responsibility for particular processes and technologies within the organization. When their responsibilities overlap, which is often the case, there is a need for greater cooperation and coordination between the departments and their end users.

On the other hand, the product design department is responsible for innovation to products within an organization. These changes can be incremental or radical: creating new add-ons for a particular product or new additions to a product family. As with the two departments shown

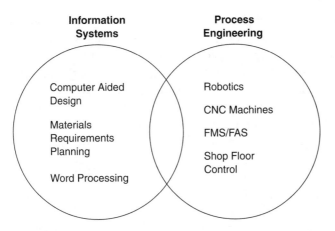

Figure 12.2   Information Systems and Engineering Departments

in Figure 12.2, there is often an overlap with the process engineering department in order to optimize design features with process capability. Other departments that involve a high degree of potential innovation include departments of quality, health and safety, environment, and so on. It must be emphasized that although some departments have high degrees of apparent responsibility for innovation and others focus more on operations, each department should be encouraged to have innovation plans in place that challenge the status quo. Each department can have its own goals, actions, teams, and results. The objective of an innovative organization is to have all departments actively contributing to the innovation process as opposed to a specialist few.

Types of Structures

The particular structure of departments adopted by the organization can vary significantly and influences its ability to innovate successfully. Organizations can either have a mechanistic or organic structure. Three key issues relating to the type of organizational structure are formalization, standardization, and centralization (Schilling, 2006).

FORMALIZATION

This is the degree to which the structure promotes rules, routines, and written documentation to control the behavior and practices of individuals and teams. A high degree of formalization can facilitate the smooth operation of the organization because everything is defined and documented. Formalizing the operations of the innovation process can educate people as to how the process works and allow them to understand how to promote innovative actions. Similarly, formalization can facilitate consistent decision making by managers, with particular reference to progressing actions through the various stage gates. However, high levels of formalization can also be detrimental to organizational innovation because this can lead to structures becoming increasingly rigid and bureaucratic and may prevent people from taking initiative.

STANDARDIZATION

This is similar to formalization in that the organization promotes a defined, uniform way of operating. Though beneficial for reducing variations in output, it can significantly limit innovation. Standardization promotes the idea of "one right way." This may constrain individual creativity by reducing room to question and change the way operations happen; therefore, the innovative capability of the organization may be affected.

CENTRALIZATION

The last issue is the level of centralization promoted by the organizational structure. Excessive centralization limits decision-making abilities to a small number of key people at the top level of the organization. This type of structure can lead to delays because permission for actions must be sought at every juncture, and it can also influence the types of actions that are approved. Centralization can impede innovation by locating responsibility for innovation with a small number of people and excluding the rest of the organization from the process. This can result in creative myopia.

Each of these three issues influences the organizational structure and subsequently the innovative capability of the organization. Although a mechanistic structure (characterized by high degrees of formalization, standardization, and centralization) can result in high operational efficiency, it can significantly limit the scope for creativity and experimentation. On the other hand, organic structures (characterized by low degrees of formalization, standardization, and centralization) can better nurture innovation by being more agile and adaptive, empowering people, and reacting to the emerging environment. Organic structures support creativity and improvisation, but they can also decrease the efficiency of the innovation process because they often lack the necessary formalization and standardization. Each organization must make a compromise between mechanistic and organic structures in order to nurture innovation and still ensure that the innovation process is operating efficiently.

Defining Teams

As the external environment has become more complex and demanding, organizations have sought to transform their structure toward the organic end of the spectrum in order to innovate. Consequently, organizations have shifted toward flatter structures that embrace the power of teams in order to counteract the loss of control provided by the functional hierarchy and to improve communication across boundaries. A team is group of collaborating people with a common purpose. Common purpose can be articulated through goals and actions. A group can be permanent (e.g., a department) or nonpermanent (e.g., a project team). Katzenback and Smith (1993) define a team by the presence of the following characteristics in a group:

- A small number of members to allow effective cohesion, communication, and commitment
- Mutual accountability between members for the achievement of objectives
- Members share a sense of common purpose in relation to objectives
- Defined performance objectives are associated with team members' efforts
- The presence of a common approach to the achievement of objectives

Organizations have openly embraced teams in order to increase the level of effort exerted by individual employees, to increase the correlation between task performance and goals, and to ensure adequate skills and knowledge availability to achieve complex tasks. Teams often outperform individual efforts as the effort and expertise of members is combined through shared goals. The advantage of operating in teams comes from bringing together different skill sets to respond to complex challenges (Scholtes, Joiner, & Streibel, 1996), establish clear approaches and communication structures, support real-time problem solving, and provide a social dimension that enhances commitment and performance (Katzenback & Smith, 1993). Ultimately, teams encourage flexibility, efficiency, knowledge sharing, and employee involvement that all contribute to innovation.

There are two types of teams: permanent and nonpermanent. Management teams, departments, and committees are examples of permanent teams. Permanent teams don't change significantly over a long period. Permanent teams typically define the goals and associated portfolio of projects or initiatives within the innovation process. Project teams, on the other hand, are examples of nonpermanent teams. Project teams typically define the objectives to be achieved and the associated tasks and deliverables that will be undertaken. They have a start date and an end date, defined by the project to be executed. Each time a nonpermanent team forms to execute a particular action, it can have an entirely new set of goals and team members. As we learned earlier, each new project or action is unique—it has never been done before. Therefore, a nonpermanent project team needs to define and understand a new set of goals each time a new project is started and must also learn how to work together to achieve these goals.

Innovation Teams

Within the innovation process in particular, there are a number of types of teams: senior management teams, innovation teams, action or project teams, and creative teams.

SENIOR MANAGEMENT TEAMS

The senior management of an organization can behave as a team in determining the strategic goals that the innovation process will strive to achieve. The use of a team approach in relation to decision making about particular innovative actions can increase the suitability of the organization's portfolio of innovations. Management teams typically are cross-functional, with managers from each function in the organization engaged.

INNOVATION TEAMS

Throughout this book, groups of people who share common goals and collectively engage in actions are called innovation teams. This type of team can be made up of people from one department, but such teams can also contain specialists and senior stakeholders from other functions. An innovation team is a manageable group of people that represent the organization, whether that organization is a large multinational corporation, a business unit, or a department.

PROJECT TEAMS

It is possible for people operating in different functions to form teams in order to develop and implement projects for innovative products, processes, and services. Operating as a team not only provides the correct mix of skills to meet the challenges of an innovative action but also enhances the commitment, mutual accountability and loyalty of team members so as to achieve the desired outcome. Operating under a team structure also enhances the knowledge sharing, understanding, and education of participants. There are various kinds of nonpermanent project teams, and these will be discussed later in this chapter.

CREATIVE TEAMS

Problem solving and idea generation are rich environments for the application of teams. Given the ambiguous and unstructured nature of the creative process, people with diverse backgrounds, skills, and insights operating as a team can increase the collective ability to analyze, generate, and test concepts that may lead to future innovations. The presence of differing mindsets can expose members to new perspectives that can support the creative thought process. These enhanced problem-solving and idea generation capabilities are beneficial not only at the "fuzzy front end" of the process but also as innovative actions progress to completion. This engagement of members in creative teams increases the number of concepts submitted to the innovation process and also increases the individual team members' commitment to ongoing innovation by making them aware of the importance of their own roles in the process. Creative teams can be put to work on proposing ideas that solve difficult problems or achieve particular goals. Their role is to create ideas that may later be assigned to project teams.

EXAMPLE: The organization you have created as part of your activities can be viewed as a permanent innovation team. You may have chosen to bring together people such as senior managers in a large organization or members of a particular department. You may also have simply brought together most of the participants in the various actions and projects engaged in your innovation plan. These individual actions (ideas, problems, and projects) are examples of nonpermanent project teams. The people assigned to lead and execute each action will eventually leave the team when the action has been completed.

Creating Effective Teams

An effective team has well-defined goals in which members share a common ownership. Mutual accountability, trust, and conflict resolution are all qualities of an effective team. These teams will have team leaders who are capable of motivating and engaging members and who will resolve conflicts and keep the focus on common goals. Effective teams are ones whose members experience equitable rewards and appropriate recognition for their actions and possess the necessary skills, knowledge, experience, and organizational power to accomplish their tasks successfully. The selection of team members for any project should be driven by the needs of the task at hand. Before the team is created, the organization should be clear as to its objectives; only when these are determined can the necessary competencies be elicited and the appropriate people selected for the team. By possessing the necessary attributes, the team members will have greater likelihood of achieving successful objectives. Another important reason for clearly defining the objectives is that all members will have common goals and a greater sense of commitment to the project, thereby increasing the level of accountability and loyalty within the team. A number of factors affect the operational performance of a team. These include team size, team commitment, and a supportive environment (Hackman, 2002; Katzenback & Smith, 1993).

TEAM SIZE

The size of the team influences the team's effectiveness; although more members can enhance its skills and knowledge base, they can also limit its ability to operate in harmony. Team members do share common objectives, but as numbers increase, so does the difficulty in achieving consensus. The average size of innovation teams in U.S. organizations to ensure cohesiveness and focus is estimated at 11 (Schilling, 2006). Large numbers of team members can lead to reduced communication and coordination and can ultimately affect team loyalty and the commitment necessary to ensure high performance.

TEAM COMMITMENT

This attribute is clearly beneficial for the organization because it will lead to increased motivation in achieving objectives in the face of difficult circumstances. The three categories of team commitment are affective, continuance, and normative commitment (Meyer & Allen, 1990).

- Affective commitment is the manner by which people identify with their organization or its objectives, which can result in strong emotional attachment between them. These levels are affected by the characteristics of the person and the structure of the operating environment.

- Continuance commitment can be defined as a person's need to work within the organization for financial gain and reward and marks a more calculative type of reasoning.

- Normative commitment can be defined as commitment influenced by society or the manner in which people perceive that they should relate to their organization. This perception is based on what people believe to be a fair and equitable relationship within the organization. In the team context, normative commitment is influenced by the norms of the particular team and the level of mutual accountability between individuals. This is often called team loyalty.

People can experience each of these categories of commitment in relation to both the organization and their involvement with their team. Whatever type of commitment the person possesses affects his or her behavior and motivation. Affective and normative commitments are the most desirable in a team because members are engaging intuitively out of loyalty and a sense of purpose rather than for financial reasons.

SUPPORTIVE ENVIRONMENT

Although a team can have its own norms and practices, it is also affected by the environment in which it is situated. As individual members are drawn from the larger organizational culture, they will bring certain beliefs and routines with them into the team. Similarly, the team will be required to interact with the larger organization during the fulfillment of its objectives in order to obtain resources, information, and support. Therefore, the level of support provided by the organization will influence the team's ability to achieve its goals. The presence of a supportive organizational environment can include elements such as the following:

- Senior management championing the team
- Open sharing of knowledge across boundaries
- Provision of adequate resources
- Acceptance that teams operate differently from common operational tasks

The organization's managers also need to ensure an effective alignment between individuals' personal goals, team goals, organizational goals, and reward systems. Only if team members feel that they can advance their careers and will be rewarded for their efforts in achieving the team goals will they commit fully to the task. If the organizational support structures are aligned with the requirements of the team, greater team performance can be achieved, and this increases the likelihood that the output will contribute to moving the organization along the correct path.

Project Team Structure

The structure and behavior of the individual project teams are an important part of the innovation process. The structure of the team can represent a wide cross-section of functions and disciplines in the organization. Multiple disciplines are necessary because no single type of skill provides the systemic knowledge necessary to implement the wide variety of actions that will be executed. However, multiple disciplines and a wide variety of skills can increase the size of teams to a point where they become difficult to manage. There are many different ways in which a team can be structured to execute an action or project. The type of team depends on the scope and definition of the task. For example, a project that involves the replacement of a computer system may be populated exclusively by computer specialists from one function or department. On the other hand, if the task is to develop a new product or business process, then the team will typically involve managers and specialists from a number of functions. There are four general types of project teams (Wheelwright & Clark, 1992): functional teams, lightweight teams, heavyweight teams, and autonomous or tiger teams (Figure 12.3).

FUNCTIONAL

This type of project team is typically populated by members from the same function or department. If a number of functions are involved in a project, then a specialist team would be set up for each function. Coordination between functional teams would be facilitated by the next highest management tier in the organizational hierarchy. This structure encourages projects to be divided into distinct phases in which one department hands over results of a project to the next one in order to

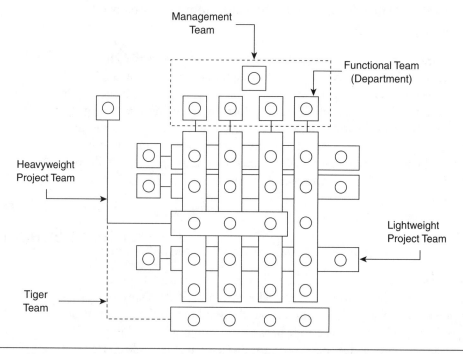

Figure 12.3   Types of Project Teams

complete the work. This is often called an over-the-wall approach because each department focuses on achieving its own individual results; the danger is that departments may have little interest in the outcome once their work is finished and the project has been passed on to the next function. Functional organization teams are typically deployed where a project has a significant need from one function and tasks that can be clearly delineated between specialists, such as installing a new computer server.

LIGHTWEIGHT

This project team has a project manager who coordinates the activities of people from different functions. The level of authority of the project manager is low. The main decision-making power regarding activities of team members still resides with functional managers. The project manager is often a functional specialist (e.g., an engineer or systems analyst), and the project has a significant functional requirement, which the specialist will coordinate and guide toward a successful conclusion. A key weakness of this structure lies in the achievement of an appropriate balance between the demands of the function and those of the project. Because control of team members rests primarily with the functional manager, the lightweight manager often struggles to secure the promised resources necessary to fulfill project objectives. Often the only way to access the necessary resources is through communication with the functional manager, extolling the benefits of the project for the particular department. Lightweight teams typically are deployed for projects that have a large requirement from one function but also need a certain level of participation from other functions, such as installing a new computer hardware system in the sales and production offices.

HEAVYWEIGHT

In this structure the team's project manager has significant power and authority in leading a significant project for the organization. Members of the team are partially released from their functions to the project and report directly to the project manager. This manager usually has senior status in the organization and has the appropriate experience and skills for decision making in order to move the project forward. Heavyweight project teams typically are deployed for projects that need significant input from a number of functions, such as designing a new process for the sales office that includes computer systems, application software, training, and so on. The benefit of this structure is that the project manager has control over team resources in order to achieve project objectives. However, the challenges associated with this structure are that it can cause conflict between the functional and project managers in achieving their own objectives. It can also significantly increase pressure on team members as they deal with work overload as a result of the dual focus of the matrix organization, and can weaken day-to-day operations as a consequence of reduced available resources. Yet if these weaknesses are addressed, this structure ensures that a cross-functional team with the appropriate resources will facilitate achievement of project objectives. This type of team structure is used for projects that have ambitious targets requiring significant investment and have the potential to result in radical innovation. The inconvenience of the structure can often be deemed worthwhile, given the potential outcome and the fact that it can lead to the likelihood of success for projects.

AUTONOMOUS

This type of structure ensures that team members are removed from their functions and dedicated exclusively to the project manager for the duration of the project. This team typically resides in a common location, allowing effective communication and complete dedication to the goals of the project. Autonomous organization teams are deployed for projects that have a major strategic significance for the organization and need a large investment of time, money, and labor. They are also used when the value of speedy attainment of objectives far outweighs the disadvantage of dedicating people solely to the project, with the associated costs and disruptions this entails. An example of such a project is the design and installation of a new manufacturing facility for a new product that differs substantially from the existing product family. The first benefit of an autonomous team structure is that it can focus exclusively on achieving project objectives without having to balance day-to-day operational commitments. Second, because the organization is dedicating people to the project for its duration, they will probably have greater support and access to skills and funding. As the team is self-managed, it can use its initiative in achieving objectives and thus can often deliver radical breakthroughs. The weaknesses associated with this structure are that as team members are released from their normal duties, they can lose touch with the emerging needs of the organization or can leave the rest of the organization in the dark as to ongoing project developments. Additionally, moving people from their usual workplaces not only is costly for the organization but can also lead to a knowledge vacuum that reduces day-to-day operational efficiency. Despite these weaknesses, the autonomous structure results in a dedicated team that can commit completely to achieving project objectives and the delivery of innovation for the organization.

Ambidextrous Structures

Each of the project team structures discussed in the last section is suited to the requirements of particular projects. The pursuit of incremental innovation or the more demanding and longer-term radical innovation places different demands on the organizational structure (O'Reilly & Tushman, 2004). Not every project can attract the resources of a heavyweight project manager. Heavyweight project manager structures are important when the impact of the project crosses a number of functional boundaries and when the project is strategically important for the organization. Autonomous teams are used for extremely important, resource-intensive, high-risk projects, typically those that involve disruptions to current business practice. Whereas incremental innovations can be undertaken within the traditional structures of the organization, the more breakthrough innovations often require significant structural adaptation such as cross-functional and autonomous teams. In a study of 35 organizations engaged in both incremental and radical innovation, O'Reilly and Tushman (2004) found that organizations launching breakthrough innovations were significantly more successful with an ambidextrous structure. They identify this type of structure as one where the organization separates its radical exploratory units from its established exploitive one (similar to autonomous teams) but still ensures that the senior management team maintains strong integrative links over both structures. Such an ambidextrous organization allows managers to balance and coordinate the competing trajectories of incremental and radical innovation. The structure allows the radical focused units to benefit from sharing the resources of the traditional units without being constrained by established culture, practices, and structures.

EXAMPLE: Over its history, Lockheed Martin Aeronautics has produced a number of innovative aircraft that have broken existing aeronautics standards. One of the most innovative of these was the SR-71 Blackbird, developed to fly high-altitude spy missions over the Soviet Union. Commissioned in the early 1960s, a long-range reconnaissance plane capable of altitudes of 80,000 feet and speeds of Mach 3 was deemed impossible by some industry experts at the time. Yet when Lockheed won the contract, they compiled a dedicated team of their best people to make the Blackbird project a reality. This team was located at the legendary Lockheed Skunk Works facility, where they had a high degree of autonomy and access to funding and technology to facilitate their efforts. The designers, engineers, and other specialists were all located close to the shop floor, and an approach of prototyping was adopted to shorten the development cycle. Less than 2 years after accepting the contract, the Advanced Developments Team delivered the revolutionary prototype in late 1964, and, after testing, the plane entered military operations in early 1966. The technological superiority of the SR-71 and the success of its development team are highlighted by the fact that the Blackbird line remained in operation until the late 1990s.

Team Empowerment

Empowerment occurs when senior managers give lower organizational levels the necessary power to make decisions on their own. This power provides teams with autonomy in decision making and encourages responsibility in directing their own activities; this not only speeds up operations but also increases the team's ownership of the particular action. An empowered team member can act on his or her own initiative, experiment with potential solutions, and react faster to emerging circumstances. Six psychological job criteria are related to an empowered workforce (Pava, 1983):

- Autonomy and discretion
- Sense of meaningful contribution
- Opportunity to learn and continue learning on the job
- Optimal variety
- Opportunity to exchange help and respect
- Prospect of a meaningful future

Individuals and teams seek autonomy and discretion in the goals they pursue and the actions they complete that take them toward their goals. These goals must be balanced with the overall goals of the organization. Individuals seek a sense that their efforts are meaningful in achieving the overall goals of the organization. They also seek an opportunity to continuously learn and develop in their careers. Individuals and teams seek optimal variety in their tasks that will put their skills to work and offer new challenges appropriate to their skills. Individuals and teams also seek opportunities to exchange help with others and to have their efforts respected in an atmosphere of parity of esteem with colleagues. Finally, individuals seek the prospect of a meaningful future in the organization that values and respects their contribution and potential contribution in the future.

Empowerment and Enablement

One of the key issues to be addressed in any team formation is who should be allowed to participate in particular innovative actions and in particular tasks. Team success depends on achieving the correct balance between empowerment and enablement. Enablement is the ability of the team members to execute the tasks of the project, and it represents the skills and knowledge of each person on the team. Empowerment is the amount of authority and discretion the team has to make the necessary decisions to execute project activities. There are four possible scenarios in the balance between empowerment and enablement: low empowerment and low enablement, high empowerment and low enablement, low empowerment and high enablement, and high empowerment and high enablement.

An individual team member reaches his or her full potential only when high enablement and high empowerment are achieved. Every individual and team in the organization must strive for this position regardless of the task at hand. Someone responsible for processing an invoice is as entitled to achieve optimum empowerment and optimum enablement as a senior manager. A key goal for every organization is to reach the highest level of empowerment and enablement for each individual and team that is appropriate for the task at hand. The organization can increase enablement of individuals through training and experience by participation in teams on other projects.

Team Skills

The level of enablement needed by team members is determined by the task at hand. The team composition does not require that each individual member possess similar skills but that they collectively bring together the necessary skills to achieve their task. Skills can be divided into four categories (Luecke, 2004): technical, creative and problem solving, interpersonal, and organizational.

TECHNICAL SKILLS

These refer to specific expertise that is necessary to achieve the task objectives. For example, if the team is charged with innovating their process by developing an enterprise resource planning system, then they will need team members who have skills in software programming and system integration. There are a wide variety of technical skills. A comprehensive view of skills might be achieved by reviewing all the individual courses, diplomas, and degrees available at a number of colleges, universities, and training organizations and selecting those appropriate to the organization.

CREATIVE SKILLS

If a team is charged with problem-solving tasks, then its members will need creative skills that allow them to analyze, incubate, and develop solutions to the challenges they will encounter. Problem solving and creative thinking skills are enhanced through the use of tools and techniques and through experience. Creative skills are transferable between functions and even careers. Engineers may learn problem-solving skills within engineering that are easily transferred to other domains, such as managing a department or organizing a charity event. Creative skills include problem-solving skills, idea generation skills, project management skills, and all of the innovation management skills discussed throughout this book.

INTERPERSONAL SKILLS

Operating as a team requires people to share common goals, to collaborate, and to be mutually accountable for the achievement of the team objectives. Therefore, the presence of skills that facilitate social interaction within the team is important. These types of skills include communication skills, knowledge of social structures, and a strong sense of diplomacy. Strong interpersonal skills can improve the cohesion of the team, increase commitment and motivation, and reduce conflict between members, which in turn increases the probability of a successful outcome. Such skills include the ability to articulate ideas, coaching, delegation, listening, risk taking, mentoring, performance monitoring, motivation, personal integrity, problem solving, conflict management, networking, self-awareness, and self-development.

ORGANIZATIONAL SKILLS

Because projects can be large and complex, team members must possess appropriate organizational skills in planning and managing tasks and facilitating interaction in the larger political context. Without such skills, there may be inefficient use of resources and conflict between the team and the rest of the organization. Such skills include goal management, project management, team management, result management, collaboration, networking, and a wide range of management skills such as leading, planning, controlling, staffing, and organizing.

Although it is ideal if all team members possess high skill levels across these four categories, they rarely do in reality. It is the responsibility of the team leader to coordinate the activities of the team members, be cognizant of their various strengths and weaknesses, and point out opportunities for attaining new skills in the future through in-house or external courses, workshops, and conferences. Although diversity can be beneficial for creativity, it can also result in conflict if not managed carefully. Team leaders must manage the interaction of team members and ensure that systems are in place to address difficulties before they harm the team dynamic. The organization must allow time for the team to develop through the phases of forming, storming, and norming (Tuckman, 1965). The more time the team spends together, the more they learn from each other and the more efficiently they work together. The team environment is a learning one in which people can develop skills in areas where they are weak. Participating in a team can increase enablement and allow more engagement in achieving common goals.

EXAMPLE: The skills or training programs adopted by SwitchIt Corporation (a manufacturing division) for their innovation teams are illustrated in Table 12.1. One new skill has been added this year—delegating to others—and a customized course for this is being developed by a subcontractor.

The relationships between skills and people on the team are illustrated in Table 12.2. The dark-shaded cells indicate courses competed. The light-shaded cells indicate that a course is planned.

From this the company can identify the current and future skills of each of their employees. They will also be able to identify gaps where they need to increase their employee skills to allow them to innovate in the future.

Table 12.1   Skills at SwitchIt (Partial List)

Skills	
Group	**Title**
Personal	Managing time
Personal	Negotiating skills
Personal	Communication and presentation
Personal	Project management
Interpersonal	Managing conflict
Management	Innovation management
Personal	Leadership
Interpersonal	Delegating to others
Management	Monitoring performance
. . .	. . .

Table 12.2   Skills Versus Individuals (Partial List)

Relationships								
	Skills							
Individuals	Managing time	Negotiating skills	Communication and presentation	Project management	Managing conflict	Innovation management	Leadership	...
Andrew Kelly	▓					▓		
Brenda Mooney		▓		▓		▓		
Danny Mulryan		▓		▓	▓	▓		
David Noone	▓	▓		▓				
Gary O'Halloran		▓		▓				
James Fogarty	▓	▓	▓	▓				
John Sheehan		▓						
Mary Roche					▓			
...								

Virtual Teams

Virtual teams are teams of people who work together on common objectives but are not co-located and perhaps rarely meet face-to-face. These can also be people who are co-located but whose practice reflects that of traditional virtual teams through the use of groupware technology. The virtual team relies mainly on mobile and online communication services to collaborate and share information. As with virtual organizations, these are becoming more common because of increasing globalization, the high rates of organizational change, and the availability of ever more sophisticated communication technologies.

Virtual team = Team + Groupware + Communication technology

Mobile and online communication means that the team can work together from anywhere in the world and even at different times of day. Traditional teams usually are co-located in the same building for the same working hours and share the same language and culture. Virtual teams dissolve these boundaries out of necessity, by using such technologies as online collaborative software, mobile computing, and videoconferencing. Even in virtual teams, it is important not to eliminate face-to-face contact altogether because this physical interaction builds trust and understanding between members. Virtual teams need fast and effective communication technology. There are a number of basic characteristics of virtual teams (Johnson, Heimann, & O'Neill, 2001): increased flexibility, improved interaction, dispersed project responsibility, and increased communication.

INCREASED FLEXIBILITY

This can affect how people spend their day and how they accomplish their work tasks. Their working hours generally are not strict as long as the job is finished satisfactorily and schedules are adhered to. The pressure and stress associated with working in a fixed location and following a strict timetable are reduced by the possibility of taking frequent breaks while engaging in difficult tasks. This flexibility will lead to the reduction of stress and a better job environment.

IMPROVED INTERACTION

Virtual teams increase participation levels. The virtual team allows people to interact regardless of their geographic location, which makes it attractive to new members. With the increased use of teleconferencing, videoconferencing, and collaborative portals, people will feel encouraged to join a team regardless of their location. Those on the move can continue to join important meetings by using phone conferencing or collaborative software. It is no longer necessary to miss meetings while traveling. Virtual teams mean that anyone anywhere can be chosen for a particular project. This increases the possibility of bringing new skills and new knowledge into the team.

DISPERSED PROJECT RESPONSIBILITY

Virtual teams mean that change in one location can be allocated to a team of people in other locations. Organizations can focus more on the outcome of collaboration than on the number of hours worked or meetings held. Virtual teams increase the amount of autonomy and discretion for their members, which can lead to an improved sense of responsibility and ultimately shorter response times.

INCREASED COMMUNICATION

Communication is often increased rather than reduced through virtual team technologies and processes. Collaboration between people from various parts of the extended organization increases the flow of new knowledge and ideas.

Innovation teams can be virtual, and all teams, whether virtual or not, can adopt many of the technologies that support virtual teams, such as collaboration software, which includes content management systems and data mining; teleconferencing and videoconferencing; and mobile technologies, including voice and text messaging. These can all improve communication, availability, and flexibility in achieving organizational goals.

Although many advantages are associated with the adoption of virtual teams, these can also pose new challenges for the organization. The main one involves satisfying the need for sharing, collaborating, and exchanging information. Some key requirements must be fulfilled so that successful implementation of the virtual team can be realized. These requirements include the following:

- Greater collaboration, such as coauthoring, document version control, application development, group editing and reviewing, and report annotation
- More communication, such as sending notes, announcements, and notices; talking; faxing; virtual meetings; commentary; and interactive discussions
- Improved coordination, including online project management tools, corporate calendars, group meeting schedulers, tracking communication activities, and group documents
- More training in interpersonal communications, cultural diversity, and writing of e-mails, reports, and so on

Communities of Practice

Organizations typically structure themselves around departments and various kinds of teams, both permanent and nonpermanent. These innovation teams and project teams implement goals through a portfolio of projects and tasks. Another type of organization that can be formed but that has a looser sense of common purpose is a community of practice. This supports people who are working on common tasks or are in the same professional field and who can benefit from sharing knowledge and experience. Membership usually is voluntary. These teams do not usually have specific deliverables for their interaction but instead focus on shared learning.

Communities of practice are groups of people who share a concern, a set of problems, or a passion about a topic and who deepen their knowledge and expertise by interaction. They facilitate innovation by generating new knowledge and sources of ideas. Another aspect of this definition is that they are not locked within traditional organizational boundaries because they can extend past these boundaries into the extended organization and beyond. They often develop on their own and may prosper whether or not they are a formal part of the organization. Communities of practice take many forms, including the following (Wenger, McDermott, & Snyder, 2002):

- *Small or big:* Some are small and very exclusive, involving only a few specialists, whereas others consist of thousands of people.

- *Long-lived or short-lived:* This varies based on the topic involved.

- *Co-located or distributed:* Some communities are initiated among people who work at the same location; others are distributed over wide areas. The explosion of new technologies and the need for globalization have made distributed communities of practice very popular.

- *Homogeneous or heterogeneous:* Some communities are homogeneous (i.e., composed of people from the same discipline and background), whereas others are composed of people from many different backgrounds.

- *Inside and across boundaries:* Communities of practice can exist within an organization but also across organizational boundaries.

- *Spontaneous or intentional:* Many communities start without any intervention or effort from the organization, whereas others are intentionally developed by the organization.

- *Unrecognized or institutionalized:* These can evolve through five stages. (1) Unrecognized: These may be invisible in the organization and sometimes even to members themselves. (2) Bootlegged: These are visible only informally to a circle of people in the know. (3) Legitimized: These are officially sanctioned as a valuable entity. (4) Supported: These are provided with direct resources from the organization. (5) Institutionalized: These are given an official status and function within the organization.

Summary

Each organization consists of permanent and nonpermanent teams. Project teams for implementing innovative actions are nonpermanent. Teams have a number of attributes, but in particular they are a group of people with a common purpose. Effective teams depend on a number of factors, such as effective communication, empowerment, enablement, and structure. The application of teams allows the organization to better adapt to the challenges of its external environment by gaining a more cohesive skill set and higher commitment. Innovative organizations have policies and procedures in place that support the development and training of teams. In the final chapter of this part of the book, we will examine how an organization might motivate and reward people to engage in teams and contribute to the innovation process. By doing this, the organization will have greater resources to facilitate the innovation process, both in generating ideas and, as we shall see, in fostering the necessary skills for development.

Activities

This activity requires you to create a list of skills used or needed by people in your organization. These skills can be technical, managerial, personal, interpersonal, or organizational. Skills may be articulated as a list of potential training courses that the organization has deemed advantageous for employees to complete. As new projects are added to the portfolio that require new skills, new training programs should be added to the list. Copy Table 12.3 into a spreadsheet and complete the fields.

STRETCH: Other elements to this activity may include creating a list of all people in the organization and detailing the skill set of each, which the organization can draw on to advance innovative actions. Similarly, identify the skill gaps for each employee where increased training will enhance the organization's innovative capability. Create a list of potential training courses available to increase skill levels in your team. Visit the Web pages of local colleges and training centers and make a list of potential courses. Highlight how each new course will add to the organization's absorptive capacity and better enable it to innovate.

REFLECTIONS

- What is a team, and how does it differ from an organization or community?
- What can organizations do to create a successful team environment?

Table 12.3   Create Skills

Skills	
Group	**Title**
Technical	
Technical	
Technical	
Technical	
Personal	
Personal	
Personal	
Personal	
Personal	
Interpersonal	
Interpersonal	
Interpersonal	
Interpersonal	
Management	
Management	
Management	

Group: Label of the skill (e.g., "Technical," "Personal," "Interpersonal")
Title: The skill or course in less than 12 words

- Name the four different types of teams that can be used for managing projects.
- Explain the terms *fully empowered* and *enabled* in regard to team members.
- List a number of ways in which creativity can be improved in teams.
- What technologies can be used to support virtual teams?
- What is a community of practice?

13 Motivating Performance

P eople make innovation happen. They generate creative ideas and solutions, they undertake experimentation and design, and they implement projects that bring creative concepts to reality. The more people engaged in an organization's innovation process, the greater its innovative capability. Motivating the behavior of people by encouraging them to engage in the innovation process is crucial to any organization's innovative practice. One of the most practical ways to do this is to link personal performance with organizational performance. The performance of the organization is articulated through such goals as strategic objectives, performance indicators, and stakeholder requirements. The personal success of an individual in achieving these goals can act both as a motivator and as a mechanism for performance appraisal. Every organization has a right to expect that the people it employs will contribute toward organizational goals. There are many intrinsic and extrinsic means of motivating individual performance. Intrinsic techniques such as bestowing autonomy and discretion are generally deemed the most productive. Extrinsic techniques such as linking performance to pay and other rewards may also be necessary incentives to take people out of their comfort zone so they will be engaged in positive risk taking. This chapter looks at the importance of individual motivation, the techniques that influence the efforts of individuals and teams, and the associated role of training and development in supporting innovation.

LEARNING TARGETS

When you have completed this chapter you will be able to

- Examine the role of motivation in influencing individual behavior
- Understand the difference between intrinsic and extrinsic motivation

- Outline the process of linking individual performance to organizational performance

- Describe a number of techniques for sharing rewards with all team members

- Understand the performance appraisal technique

- Explain the training and development approach in relation to motivation

- Design a simple performance appraisal system

Motivation

Individual motivation is clearly essential for achieving organizational objectives. High motivation results in increased commitment by the workforce and less resistance to change. Maslow (1954) presents a widely accepted and very succinct model of individual motivation called the hierarchy of human needs. The hierarchy divides human needs into five areas, listed here in descending order of attainment (i.e., the first is the most difficult to attain): self-fulfillment, ego or esteem, social, safety, and physiological needs.

Maslow has determined that needs become motivators only when they are left unattained. When a lower-level need is satisfied, the person experiences the need to satisfy the next higher-level need in the hierarchy. For example, typical physiological needs include food and water; we need them to live. Once the hunger need is satisfied, then our needs change to safety needs; for example, we need a safe place to live. When both these needs are satisfied, they are replaced by social needs (e.g., the need for social interaction and companionship), and so the movement up the hierarchy continues. It is important to understand that a person will be motivated by fulfillment of a higher-level need only if his or her lower-level needs have already been satisfied. Many people in an organization have met their lower-level needs and are focusing on self-fulfillment needs. These include the feeling that our personal, interpersonal, and professional lives are fulfilled. Maslow's insights touch on a number of concepts discussed earlier, including empowerment, enablement, and the psychological job criteria such as the need for optimal variety and discretion in our jobs. Organizations need to motivate people, but this involves much more than just increasing financial incentives. In order to motivate, organizations must demonstrate that they trust and value their staff and provide the freedom and empowerment to allow them to determine how they may achieve objectives. This includes making jobs more complete, interesting, and challenging; focusing on developing individual skills and career prospects; and creating an environment that drives out fear and encourages initiative (Luecke, 2006). One important factor that leads to a motivated workforce is the recruitment of the right type of people into the organization.

Recruitment must strive to attract people who are committed, who actively seek responsibility, and who are capable of self-control and self-direction in relation to achieving organizational goals. A highly motivated staff is more likely to engage fully in the innovation process.

Theories X and Y

Over the history of the organization, management perspective has shifted from the view of employees being just an input into the operation process toward the view of them being a valuable competitive asset (e.g., potential innovators). McGregor (1960) represents these two contrasting perspectives as his Theory X and Theory Y approaches to management. Theory X is an authoritarian management style that perceives the average person as being one who dislikes work, avoids responsibility, and must be controlled and directed. Theory Y is a more participative management style that views people as capable of being committed to their work and actively seeking responsibility and discretion. They are also capable of self-control and self-direction in relation to achieving organizational goals. The means by which the organization can motivate people depends on the perspective adopted by management. If management uses a Theory X perspective, then motivational tools such as the carrot-and-stick method of financial incentives and fear of punishment will be used. If management adopts a Theory Y perspective, then their motivational tools will include training and development incentives and the opportunity to fulfill self-actualization needs. Innovative organizations often adopt the Theory Y approach, where people are viewed as capable of being a competitive asset for the organization and as desiring to engage in the innovation process and contribute to organizational development.

Intrinsic and Extrinsic Motivation

Motivation drivers can be divided into two categories: extrinsic and intrinsic. Intrinsic motivation comes mainly from within the person and can be influenced by factors such as enhancing the scope and purpose of the job role in order to stimulate and satisfy intellectual capability. Extrinsic motivation comes mainly from outside the person and includes such factors as level of pay, bonuses, and other recognition initiatives. Intrinsic drivers have long been recognized as a more powerful motivational force than extrinsic drivers. Financial incentives do not motivate people to the same degree as in the past. As living standards rise, affluence and increased education have become the norm, leading to less interest in financial rewards and more in achieving job satisfaction through interesting work.

INTRINSIC MOTIVATION

This comes mainly from within the person and is evident in those who voluntarily engage in an activity for its own sake. This can happen because of an inherent desire to fulfill certain higher-level needs. Since the majority of people want to achieve higher levels of self-development and self-actualization (Theory Y), they will drive themselves to undertake new challenges rather than be content with the status quo. Intrinsic motivation is often difficult for the organization to influence because it is specific to the person's unique context. The prospect of a reward for achievement will not influence a person's intrinsic motivation. However, a suitable organizational culture can establish norms and standards within people that can create the desire to achieve higher-level needs. Similarly, the presence of supportive training and development systems can nurture the desire for specific achievement. The topic of individual development will be discussed in more detail later in the chapter. Although intrinsic motivation may be a more powerful driver in engaging people, it is difficult to use as a management approach. The forces that act against exclusively intrinsic motivation include the following:

- People are often risk averse and tend to remain within a certain comfort zone rather than stretch beyond it.

- The intrinsic needs of the person and those of the organization may not align.

- People need to fulfill all of Maslow's needs, and lower needs are not adequately met by intrinsic motivation and instead require extrinsic motivation.

EXTRINSIC MOTIVATION

This uses external rewards and recognition to influence the person's behavior in a certain manner. To counteract the forces against exclusively intrinsic motivation, organizations often use the extrinsic motivation of reward and recognition to guide and influence individual efforts. Rewards and incentives such as financial incentives, additional holidays, and other rewards are used to motivate people. Recognizing efforts either informally through praise or more formally through promotion and advancement can also provide the necessary motivation. Although extrinsic motivation is readily used to engage people, financial incentives will motivate most people only up to a certain saturation point, after which other factors such as free time and self-development will become important. The ideal scenario for many organizations is to maximize the factors that provide intrinsic motivation and then optimize the remaining requirements with an extrinsic-based system. In addition to cultural factors discussed earlier, three formal techniques can be used to provide extrinsic motivation (i.e., linking reward to performance): gain sharing, profit sharing, and performance appraisal. The third of these, performance appraisal, can also be used to enhance intrinsic motivation.

Gain Sharing

Gain sharing is a group bonus scheme in which the entire organization shares in the benefits that result from improved innovation. This system links individual remuneration to organizational rather than individual performance. When organizational growth occurs, people benefit as a whole through mainly financial bonuses. When gain is shared across the organization, there is less competition between team members. On the other hand, some people may be rewarded for very little effort. In addition, this system increases the possibility that all people may operate at a similar level of productivity (i.e., highly productive people may reduce their overall effort to the norm across the organization). Gain sharing promotes teamwork and more pragmatic reward-sharing incentives. Gain sharing uses a formula that is simple to understand and easily related to innovation, productivity, or profitability. A successful gain sharing program owes its success to the formula and education for employees on how to improve their benefits through the scheme. The formula must be based on a careful examination of the organization's past performance indicators. Factors such as productivity, quality, cost of compensation, and reduction in lead times can be used in developing this formula. One example of a gain sharing formula is, "If workers keep labor costs below 12% of sales, then they get a bonus amounting to 60% of the difference between the actual labor costs and the 12% target." Gain sharing formulas must be kept simple and easy to understand. Benefits commonly attributed to gain sharing include the following (Lawler, 1992):

- Coordination, teamwork, and sharing of knowledge increase.
- Social needs are recognized via participation and mutual reinforcing of group behavior.
- Attention is focused on goals such as performance indicators.
- Change due to technology, market, and new methods gains acceptance.

- People demand better performance from each other.
- Innovation increases.
- Where unions are present, union–management relations become more flexible.

There are four critical components for developing a successful gain sharing plan: management commitment, individual involvement, structures, and communication.

MANAGEMENT COMMITMENT

Managers must wholeheartedly support the plan. They must also work to develop a company culture of respect, open communication, and cooperation. Managers must give individuals and teams the power to adapt so that they can better engage in innovation and realize the innovation gains.

INDIVIDUAL INVOLVEMENT

Teams must have a variety of ways of sharing information, tackling problems, and monitoring results that focus on individual requirements. This includes training in personal skills, interpersonal skills, and technical or management skills.

STRUCTURES

Structures must be put in place that facilitate and encourage participation in the innovation process. The presence of idea generation, problem solving, team participation, and result monitoring systems can all encourage engagement in the innovation process. Structures can include meetings, teams, regular training, procedures, or whatever nurtures and motivates people to behave in a specific manner.

COMMUNICATION

Information that traditionally belonged to management must be shared with all people across the organization, such as news of upcoming orders or customer complaints about products. Feedback on organizational performance and internal difficulties is also an important part of any gain sharing plan because it communicates any pressing requirements on the organization and can provide a stimulus for people to develop future innovations.

Profit Sharing

Profit sharing is an incentive plan that pays bonuses based exclusively on profits as opposed to growth in productivity or market share. The plan gives people the potential for high rewards if the company thrives. Profit sharing schemes can develop a sense of ownership among employees that results in increased individual motivation, productivity, and performance. Profit sharing and gain sharing differ in that gain sharing holds people accountable for a few key expenses that they can control (e.g., material costs, output) and is generally paid out on a quarterly basis. Alternatively, profit sharing is affected by elements that cannot be controlled (market fluctuations), and the payout may be over a longer time frame. There are many variations of profit sharing plans, including the following: approved share participation plans, share subscription plans, share option plans, save-as-you-earn plans, and restricted share plans.

Each plan has its own merits. On joining the save-as-you-earn plan the employee agrees to save a fixed sum out of net pay for a predetermined period (usually 5 years). Share options are granted based on the amount saved. At the end of 5 years the employee can use the proceeds to buy some or all of the shares covered by the option, or continue to invest in order to qualify for a higher sum in the future. Share prices may be subject to sudden and often dramatic fluctuations and that are often outside the control of the particular company. A good underlying company performance reflects itself over time in increased share value. That fact, together with the variable nature of the investment, must be made known through an effective communication program. It is essential that employees completely understand what is involved in profit sharing plans.

Performance Appraisal

Management always strives to motivate people to engage in innovative actions that will contribute to the achievement of goals. One of the most valuable ways of achieving this is through the performance appraisal system. This is used to reward or recognize people through individual performance review and, where appropriate, through annual salary increases. This method structures the interaction between an employee and his or her supervisor in order to formally appraise the employee's progress on a number of goals, both organizational and personal.

The appraisal system provides management with a means of linking strategic goals to individual goals. The appraisal incentives should reinforce the organization's strategy of play-to-win (Davila et al., 2006) in order to encourage employees to challenge the status quo and behave in a creative manner. In this way, the employee can strive to contribute to the achievement of certain goals and be sure of a positive personal outcome from his or her efforts in terms of organizational reward and recognition. Consequently, many organizations use appraisal results, either directly or indirectly, to determine the rewards (both financial and promotional) that their staff should receive. The appraisal results are used to identify high achievers, who should be the first recipients of available merit pay increases, bonuses, and advancement within the organization. Similarly, appraisal results are used to identify poorer performers, who may need some form of mentoring and training to improve their future performance. There are a number of important factors to be considered when setting up an effective performance appraisal process: setting organizational goals, designing the appraisal process, linking appraisal to rewards, and rewards and team behavior.

SETTING ORGANIZATIONAL GOALS

It almost goes without saying that when an organization is setting motivation goals, the broader goals of the organization must be clearly defined and understood by all involved. Performance metrics appropriate to the goals must also be created and proper alignment created between the reward system and these defined goals and metrics. If goals do not meet these criteria, then linking them to individual performance will have little or no effect because the employee may have little understanding of how increased effort and risk taking can lead to desired rewards.

DESIGNING THE APPRAISAL PROCESS

In order for performance appraisal to work effectively, the organization should strive to ensure adequate design of the process, from issues such as the timing of appraisal and follow-up meetings to the identification and rectification of bad performance (Luecke, 2006). A crucial part of any appraisal is not only reviewing achievement of predefined goals over the previous period but also setting new objectives for the coming period. The two parties involved must reevaluate the goals being pursued to ensure that they are appropriate for the coming period. In this way, the organization ensures that individual behavior reflects the emerging needs of the organization rather than the historical ones. This facilitates the generation of innovative actions that meet the existing and emerging requirements of the organization.

LINKING APPRAISAL TO REWARDS

There is significant debate on the need to link appraisal systems with the organization's reward system. Advocates argue that doing so can lead to significant increases in innovation and productivity by linking individual goals (money, advancement, and recognition) with the desired goals of the organization. Critics argue that increases in productivity should not be linked with rewards; instead, the presence of a suitable culture can tap into a person's intrinsic motivational needs more effectively. Linking individual reward systems to the appraisal system can in certain circumstances can impede innovation by increasing competition between employees and reducing collaboration. Managers must be careful when linking appraisal with reward to ensure that they do not adversely affect the innovative culture.

REWARDS AND TEAM BEHAVIOR

A reward system based on individual performance can have a detrimental effect on team behavior. Team members will be unable to demonstrate their individual contribution to specific goal achievement and therefore may alter their behavior away from the collaborative and mutual accountability of teamwork. One compromise argues for a performance appraisal system that is linked to a group-based reward system (similar to gain sharing) in which reward is shared equally across a team. The appraisal process typically involves agreeing to specific development topics at the beginning of each year and then appraising the individual and team performance at the end of the year. Whereas the team appraisal is used to determine financial reward, the individual appraisal system can influence the development and training of individuals.

Performance Appraisal System

The performance appraisal system typically is constructed around a range of organizational and personal development topics. In the context of the approach adopted in this book, people can find themselves motivated to contribute to specific personal goals in the following areas: contribution to specific objectives, contribution to specific indicators, development of specific technical and management skills, and development of specific personal and interpersonal skills.

SPECIFIC OBJECTIVES

The person is given responsibility for a number of strategic objectives to be achieved either individually or as a team. These objectives usually are the same objectives or a subset of the main objectives held by the organization as a whole. People are usually measured on up to three of the most important strategic objectives for them as an individual.

SPECIFIC INDICATORS

The employee is given responsibility for up to three performance indicators, either individually or as a team. As mentioned earlier, these are usually the same indicators or a subset of organizational indicators. Up to three of the most important indicators are used to measure the employee's performance.

TECHNICAL AND MANAGEMENT SKILLS

The employee agrees to develop specific technical, management, and creative skills over the course of the appraisal period, such as computer programming skills or project management skills. Particular training programs will then be put in place that will add greatly to the organization's future skill base and become a foundation for future innovation. The employee can also be assigned a number of leadership skills to develop.

PERSONAL AND INTERPERSONAL SKILLS

The employee is assigned a number of personal and interpersonal skills to develop either independently or through attendance at particular courses of study or training. The list of potential skills includes idea articulation, coaching, customer orientation, delegation, listening, initiative and risk taking, mentoring, performance monitoring, motivation, problem solving, self-confidence, and teamwork.

Training and Development

The training and development of employees has already been identified as a reward mechanism for individual motivation because it enhances future advancement opportunities. It also results in greater longer-term benefit for the organization than is possible from other forms of extrinsic motivation such as financial bonuses. The training and development of employees increases their ability to undertake difficult tasks and consequently improves the output of their efforts toward organizational goals. As an employee's training and education increase, the employee's desire to fulfill higher-level motivation needs, such as self-actualization, increases and therefore the employee is intrinsically motivated to engage in more challenging activities.

Organizations can develop employees through job design, task delegation, skill training, and career development (Luecke, 2006). As a consequence of these actions, the organization's human resource skills and competences increase, as does the employee's ability to act autonomously. The decision of how individual employees are developed is a product of two factors: the skill deficit within the organization relative to existing and future challenges and the needs and existing abilities of the employees. Because training budgets are limited, employees will be given development opportunities that provide benefit to the organization first and to them as individuals later. Managers are reluctant to incur the expense of providing training for employees that will not directly benefit the organization in the short to medium term. The design of the training and development plan for most organizations begins with identifying the skills the organization will need in the future and assessing the current skills of its employees. With effective planning, the organization can ensure that it develops the missing skills in suitable employees through the training plan so that the competence is available when needed.

Organizations that continuously develop their employees create a culture of questioning, creativity, and experimentation that is beneficial to the innovation process in generating ideas and solutions that may lead to innovations. The development of employees can also provide them with the necessary skills for greater engagement in the innovation process, which will allow greater numbers of innovations to be undertaken. The key areas where ongoing development facilitates the innovation process are technical, problem solving, interpersonal, and organizational skills. Over time, as employees receive training and development in one or more of these areas, they are capable of making a greater contribution to the organization's innovation effort.

EXAMPLE: Ivan Fallon is the operations manager of Harper Sculpting, a gymnasium group specializing in health and fitness services. Tom Griffin, a team member, is the site manager for one of the gyms. They work together as part of the development team. Every member of the team goes through performance appraisal, which encourages them to achieve a number of organizational and personal development goals. The goals are agreed on by the team member and supervisor at the beginning of the year, and individual progress is assessed during the year. At the end of the year each goal is given a score that indicates its degree of completeness or the degree of engagement by the individual. Table 13.1 shows a completed performance appraisal form for Tom Griffin. Both the team member and the supervisor have signed off on the allocated scores.

Table 13.1 Performance Appraisal

Appraisal		
Employee:	Tom Griffin	
Supervisor:	Ivan Fallon	
Appraisal Period:	<09/2006>-<08/2007>	
Status:		(during period)

		SCORE (1 to 5) (end of period)
GOALS (start of period)		
Objectives:	Achieve an empowered & happy workforce	3
	Conduct improvement meetings with increased employee involvement	4
	Measure customer satisfaction	5
Indicators:	Maintenance/downtime hrs	2
	Complaints	2
	Absenteeism	3
Skills:	Encourage and motivate other members of the company	3
	Ability to work in groups	3
	Ability to communicate effectively throughout all departments	4
	Total Score: (end of period)	29

273

Summary

Motivating performance can be achieved through a combination of intrinsic and extrinsic techniques. Intrinsic reward involves creating a stimulating and motivating work environment that appeals to the employees' need for self-actualization in their careers. Extrinsic reward involves developing a system in which the employee's performance is linked to the performance of the organization, which in some cases leads to tangible rewards such as share options and pay increases. Motivation can combine traces of both intrinsic and extrinsic motivation techniques and can tie an employee's personal goals directly to organizational goals. Organizations can use their training and development plan as a motivational reward but also as a method to increase individual skills so as to cope with future unforeseen challenges.

Activities

This activity requires you to create a performance appraisal record for one of the people in your organization. Copy Table 13.2 into a spreadsheet and complete the fields. Define your employee's contribution over the appraisal period in relation to the activities of the innovation process, together with skills he or she needs to develop over the coming period.

STRETCH: Other elements of this activity may include creating a full portfolio of performance appraisal records for all of the people on your team.

REFLECTIONS

- Why is motivation so important in innovation?
- Give an example of intrinsic and extrinsic motivation.
- What is the difference between gain sharing and profit sharing?
- Explain the performance appraisal system.
- How is the need for new skills implemented in a performance appraisal system?
- How are individual goals related to organizational goals?

Table 13.2   Create a Performance Appraisal

Appraisal		
Employee:	<Individual>	
Supervisor:	<Individual>	
Appraisal Period:	<Month/Year>–<Month/Year>	
Status:		(during period)
GOALS (start of period)		**SCORE (1 to 5)** (end of period)
Objectives:	<Objective> <Objective> <Objective>	
Indicators:	<Indicator> <Indicator> <Indicator>	
Skills:	<Skill> <Skill> <Skill>	
		Total Score: (end of period)

Part V

Sharing Innovation Results

Two of the primary causes of failure in the innovation process relate to the poor sharing of results and poor communication and sense of community. Results are defined as the outcome of an effort. Earlier parts of this book discussed three types of efforts: the definition of goals, management of actions, and empowerment of teams. This part continues these three discussions but focuses on closing the loop in the innovation funnel (i.e., monitoring and sharing the information relating to goals, actions, and teams; see Figure V). This part also looks at bringing all three elements together into one innovation funnel and understanding relationships between them. One funnel represents the innovation activities of one innovation community; this can be a management team, a department, or a project team. Later in this part, we will show how an extended organization can have many funnels or communities that all share innovation-related information. Chapter 14 looks at knowledge management and the different ways of capturing and sharing information related to innovation. Chapter 15 looks at building an innovation community and, in particular, at building a simple knowledge management system for creating, editing, and sharing innovation-related information. The final chapter, Chapter 16, looks at extending the innovation process beyond a single innovation community. It looks at various ways for sharing innovation knowledge across an extended organization that may include many departments, management teams, project teams, business units, and even suppliers, distributors, and customers.

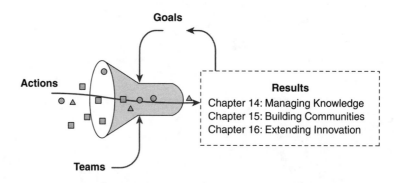

Figure V   Sharing Innovation Results

LEARNING TARGETS

When you have completed this part you will be able to

- Describe knowledge and how innovation knowledge can be managed
- Understand specific aspects of collaborative portals, portlets, and workspaces
- Discuss how to design and implement a simple knowledge management system
- Explain the relationships between different sets of innovation information
- Describe how organizations can contain a number of innovation communities
- Describe a hierarchy of innovation management processes or funnels
- Describe an approach to presenting and reporting a dynamic innovation plan

Managing Knowledge 14

Applying innovation was described earlier as offering practical tools and techniques for making beneficial changes in all types of organizations. Many of these tools were described in earlier chapters. In this chapter we look at tools and techniques for managing knowledge and, in particular, innovation information. Every tool can perhaps be described as a knowledge management tool, but in this chapter we focus on more popular tools in this field such as online portals for sharing documents and records. Knowledge is defined as information plus judgment that leads to informed decisions. Organizations use less than 20% of their knowledge. Most knowledge remains trapped inside the minds of people and increasingly in the various data stores that an organization uses, such as databases, servers, e-mail systems, personal computer files, filing cabinets, libraries, and archives. Many organizations realize that to succeed in creating sustainable growth they need to harvest and use their knowledge more effectively. They need to develop processes that increase knowledge sharing between employees and the means to store this information. This chapter looks at knowledge management and how it can be used to improve the effectiveness of innovation. We begin by looking at the definition of knowledge and knowledge management. Later we look at the different types of collaboration between people. We then look at collaborative workspaces or portals that can be used to codify and share information. We conclude the chapter by looking at the role of meetings as a common and practical way to share innovation results.

LEARNING TARGETS

When you have completed this chapter you will be able to

- Explain the concept of knowledge and knowledge management
- Understand the various stages of knowledge management

- Describe the difference between personalization and codification
- Explain the different depths of knowledge
- Explain why organizations need employees to "care why"
- Show ways in which people can collaborate in sharing knowledge
- Describe how collaboration can be enhanced through effective meetings

Defining Knowledge

Knowledge is defined as information, together with individual judgment that can lead to informed decisions. This definition does not claim that all decisions will be correct—simply that they are informed by the information that is available and made by a person with a certain ability to judge. In essence, knowledge is about making good information available to experienced individuals and groups who then make decisions based on that information. Knowledge is embodied in people, culture, procedures, routines, systems, processes, and information systems. Knowledge is usually defined in relation to two other terms: *data* and *information*. Data are words and numbers that may have some meaning to a person (e.g., words in a paragraph or musical notes on a page). When this meaning becomes clear and useful, the data are transformed into information. Information can be stored in a variety of ways and can sometimes lose the original data that gave it meaning. When we store information about something in our minds, the original data often become forgotten, but the meaning remains. For example, many cooks initially learn new dishes by reading data in a recipe. Over time they learn intuitively how to cook the same dish without any reference to precise data in the recipe (e.g., weights or quantities of ingredients).

When information is used with judgment and experience it can become knowledge. The relationship between data, information, and knowledge is illustrated in Figure 14.1. This figure suggests that we are surrounded by a large amount of data, only some of which will ever be translated into knowledge. This diagram also suggests that a large amount of data is often needed to create a small amount of knowledge; whether those data are useful or not remains to be judged. Information and knowledge must be updated continuously.

Those with information, judgment, and experience who make good decisions can be said to have reached a particular height in the knowledge hierarchy: wisdom. It is clear that knowledge requires both information and the person who judges and interprets it. The expression *knowledge management* refers to processes that facilitate the management and sharing of information that may later be translated into knowledge. In this context, perhaps the term *information management* would be a better one, but only regarding the information specifically aimed at facilitating decision making.

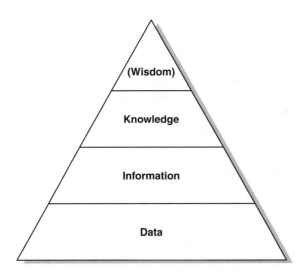

Figure 14.1   Data, Information, and Knowledge

EXAMPLE: Consider a common newspaper. It contains a large number of alphanumeric characters that are ordered in such as way as to give potential meaning to the reader; these are data. The reader typically scans these data in an attempt to discover something of value. If the newspaper is in a foreign language unknown to the reader, then it remains data. When something of value is found, the reader memorizes the data or its meaning in the form of information; then, when the reader wants to show off this knowledge in conversation, he or she exercises judgment and recalls the appropriate piece of information; this is knowledge.

Knowledge Management

Knowledge management in an organization is the process of managing information, then making it available to people so that they may exercise judgment in the decision-making process. This process can be said to consist of five stages (Bessant & Tidd, 2007): generating and acquiring knowledge, representing and codifying knowledge, storing and retrieving knowledge, sharing and distributing knowledge, and transferring and embedding knowledge.

GENERATING AND ACQUIRING KNOWLEDGE

This is a natural process in any organization. Before the startup of any organization, it is first necessary to acquire adequate knowledge about customers' requirements and demand for services or products. The activities of competitors must also be studied. Over the lifetime of an organization, knowledge is built up continuously and often shared. The depth and variety of knowledge grow over time. New knowledge can be generated through alliances and scanning of the external environment for relevant discoveries. Knowledge is also acquired through experience, experimentation, and acquisition (Bessant & Tidd, 2007).

- Experience may be the least effective attribute because it is difficult to translate into knowledge. For example, experience that is perceived rather than proven to be knowledge can lead to self-destructive habits. Consider the master craftsman whose belief in his own competencies leads him to resist all new ideas for change. This reluctance to adapt behavior to external turbulence in the marketplace may be negative for the organization.

- Experimentation is a systematic learning process leading to the acquisition of new knowledge and is a core part of the innovation process. Organizations set goals, generate ideas that meet these goals, and monitor results. If the desired outcome is achieved, then the opportunity is there to learn from its success. If the opposite occurs, then the organization needs to adapt its behavior (i.e., generate new ideas and learn from its mistakes). The laboratory is the place where the experimentation process is very clearly visible in the generation and testing of new ideas; this sometimes happens over a short time. New knowledge is generated through trial and error. The experimentation process can also be applied to the way the entire organization develops over a much longer time.

- Acquisition is another way to acquire new knowledge. Large companies acquire smaller ones, mainly for the knowledge they possess about complementary products, processes, or services (including markets). The smaller organizations bring new knowledge, and in most cases the gross knowledge of both organizations increases. However, the efficiency of knowledge transfer between the organizations may vary significantly. Alliances between organizations are another way to generate new knowledge. Collaborative research projects between companies and universities are one example of such alliances. Other techniques used by organizations to acquire new knowledge include environmental scanning for new ideas relating to products, processes, and services.

REPRESENTING AND CODIFYING KNOWLEDGE

There are two types of knowledge: tacit and explicit. Tacit or implicit knowledge resides in the minds of individuals and has not been shared or written down. An organization also possesses tacit knowledge through its norms, attitudes, culture, fears, and politics and will often have developed unwritten rules, relationships, attitudes, and procedures for achieving common objectives. Explicit knowledge, on the other hand, is acquired through written procedures or expressed through training programs. The most common form of explicit knowledge is found in the information stored in databases and other computer-based information systems. This information, when accessed, has the advantage of giving people the ability and confidence to make decisions. The option to edit this information is also available. The difference between tacit and explicit knowledge lies in the difficulty in expressing and communicating tacit knowledge and translating it into explicit knowledge.

STORING AND RETRIEVING KNOWLEDGE

Transferring knowledge from tacit to explicit involves personalization and codification. Personalization can be achieved through verbal dialog, meetings, workshops, training, and so on. Codification typically manifests itself through forms, databases, intranets, and other information management systems. Both strategies offer major challenges for an organization. Codification is now being assisted by a variety of information management tools. However, it may be difficult to motivate staff to take the trouble to codify their knowledge in order to share it. There may be a tradeoff between the difficulty of codifying and the benefits of being able to retrieve and share the information. Organizations need to pay particular attention to the design of technologies such as intranets so as to strike a good balance between the cost of codifying information and any benefits that may accrue from so doing. This important challenge is discussed later in the chapter.

SHARING AND DISTRIBUTING KNOWLEDGE

The use of computer-based communication technologies such as intranets, collaborative workspaces, and even blogs and wikis has grown significantly in recent years. There is no doubt that these technologies have contributed to an increase in the amount of new information available to all people in an organization and therefore to the organization as a whole. They have also significantly increased the amount of data that must be sifted through in order to extract information. Much of it is irrelevant and creates noise that can distract one from the useful data. Empowerment and enablement of staff to use these new technologies in order to share innovation-related information are encouraged in innovative organizations. It is also clear that the resulting access to Internet-based information has resulted in a potential increase in the tacit knowledge of individuals. The essential goal of an organization should be to adopt strategies that can maximize knowledge sharing and synthesizing different types of knowledge. However, as mentioned earlier, a balance must be struck between information that can be usefully shared and information that simply creates noise and clutters up the knowledge base of an organization.

TRANSFERRING AND EMBEDDING KNOWLEDGE

The principal way of exploiting knowledge is by setting goals and executing ideas and projects that ultimately result in adding value to customers. New knowledge can lead to radical or incremental projects that innovate products, processes, or services. This is particularly evident in the case of new product development, where knowledge may need to be protected before being shared. This final stage in the knowledge management process can lead back to the first stage, generating and acquiring knowledge, and the whole process begins again (Figure 14.2). In reality the stages may overlap, and various streams of knowledge will occur at different stages in the knowledge management process simultaneously.

Knowledge Sharing

A key objective of any organization is to share knowledge appropriately between its members, but first of all, knowledge has to be translated from tacit to explicit knowledge. Nonaka, Toyama, and Byosiere (2001) coined

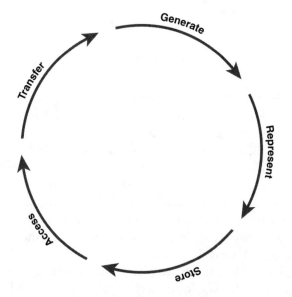

Figure 14.2   Stages of Knowledge Management

the expression *knowledge networks* to explain the way knowledge is transferred from the originating individual, to the group or organization, where it is shared and disseminated. There are four principal transfer mechanisms: socialization, externalization, combination, and internalization.

SOCIALIZATION (TACIT TO TACIT)

This involves sharing knowledge between people through formal and informal meetings and workshops and creating the environment to do it in; it includes the use of teleconferencing and e-mail. Knowledge is tacit, which means it usually resides in the minds and actions of different people. This type of transfer is sometimes called personalization.

EXTERNALIZATION (TACIT TO EXPLICIT)

This involves converting knowledge from its tacit state to an explicit state capable of being shared between people. Creating reports, slides, or even a course manual represents this type of transfer. Other examples include filling out forms for information stored in a database. This type of transfer is sometimes called codification.

COMBINATION (EXPLICIT TO EXPLICIT)

This transfer involves changing explicit knowledge from one structured format to another. For example, knowledge about customers may be found in a survey, and this may be transformed through detailed analysis into an informed report.

INTERNALIZATION (EXPLICIT TO TACIT)

This transfer involves converting explicit knowledge from reports or courses, for example, into tacit knowledge within people. This may be coupled with some form of activity in which explicit knowledge is used to carry out a task; when the task is completed, the experience and knowledge have been converted into tacit knowledge.

If all four of these mechanisms are used liberally to share collective knowledge, the knowledge base of the organization can increase significantly. This sharing of knowledge also reduces the risk of skill loss when a person moves on to another organization. The more people who are involved in problem solving and idea generation, the more the organization will benefit from the flexibility a culture of knowledge sharing will bring.

Codification and Personalization

Knowledge management often focuses on two principal strategies: personalization and codification (Hansen, Nohria, & Tierney, 1999). Personalization of knowledge involves the exchange of knowledge and experience between individuals through face-to-face meetings (both formal and ad hoc), presentations, lectures, and other information exchange forums. The emphasis in this approach is on human contact. Many organizations create physical environments where people are encouraged to do this regularly. Codification of knowledge involves using forms, databases, and other information systems as well as reports, manuals, and even presentation slides. It may be necessary to complete specific tables and data fields in order to structure particular information. This can then be stored and retrieved from the organization's information systems. Systems such as content management systems, collaborative workspaces, portals, and case-based reasoning systems are often used to store information that may be reused over and over again. The two approaches are not mutually exclusive. Codification on its own may ignore an organization's need to create culture, routines, and processes for personalization. Personalization, used on its own, may ignore the very rich source of knowledge already stored on individual computer systems, group intranets, and content management systems; these may become even richer sources if people are encouraged to codify and share knowledge.

The compromise strategy and relationship between personalization and codification of knowledge is illustrated in Figure 14.3. Three possible strategies are illustrated. The first strategy is a major increase in the codification of knowledge. This means codifying every available piece of information in the organization, with the inevitable overhead and complaints associated with keying in information that is already known. The second strategy is to dramatically increase the personalization of knowledge, allowing more time for meetings, both formal and informal, more presentations, and so on. This also has disadvantages in that it is time-consuming and may cause boredom and demoralization among those who would rather complete a task than talk about it. A compromise strategy exists in which organizations codify only critically necessary information to be shared throughout the organization as a whole. If necessary, more detailed information may be retrieved from the people involved at a later date. We call this minimum critical codification. Personalization may also be limited to specific team-based activities such as goal setting, result monitoring, and project implementation, thereby reducing unnecessary information sharing and the creation of information noise. We call this team-based personalization. Earlier you saw a number of techniques for setting up teams of common purpose and maximizing team-based personalization. In the next chapter we will look at minimum critical codification of simple knowledge management systems.

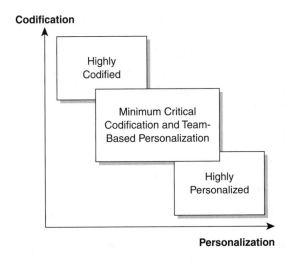

Figure 14.3   Codification and Personalization of Knowledge

Depth of Knowledge

There may be enormous variations in the amount of knowledge people possess. More experienced and educated people clearly have an advantage over newly recruited students in their depth of knowledge. Regarding a particular topic, there are four depths of knowledge a person can possess (Stewart, 1997):

- Know what
- Know how
- Know why
- Care why

A person's depth of knowledge increases from one level to the next. "Know what" is factual knowledge about a subject. Recently graduated students possess a certain level of "know what" but have gained very little experience except through work placement or college-based projects. They possess the facts but have not yet applied them to a variety of new and real-life situations. "Know how" comes from the experience of applying knowledge to a particular situation repeatedly and successfully, and usually takes a number of years to manifest itself. "Know why" is a new level of knowledge in which the person understands the broader reasons behind completing a particular task and applying a particular piece of knowledge. If they move beyond the task of installing a particular piece of equipment to understanding why this will improve the organization in some way, then they can be said to possess "know why." The final depth of knowledge, "care why," is reached when a person cares about the work environment enough to be motivated to improve and change it from within so that the broader objectives of the organization will be achieved. It is not necessary to tell such people what to do; they care enough to initiate innovation themselves.

People in the organization will have varying depths of knowledge, although the ideal situation would be to have as many as possible at the "care why" level. However, regardless of what depth of knowledge they possess, each person can play a role in innovation (Figure 14.4). A senior manager would be expected to reach a level of "care why" in depth of knowledge. At the other end of the scale, new recruits, freshly graduated from college, can be said to possess no more than a "know what" depth of knowledge; however, they can initiate an idea. This may have come from a stimulus such as a newspaper article that sparks off an innovative thought. This idea can in turn be processed by an investigator with a "know how" depth of knowledge (i.e., someone with experience in the topic in question). The final decision about the implementation of the idea can be made by the manager. In this way, all people in an organization can be involved in the innovation process simultaneously.

Collaboration

People transfer knowledge in a number of ways. Earlier in this chapter, we saw how tacit and explicit knowledge can be transferred. In this section we look at another perspective on ways in which knowledge can be transferred, focusing on the time and place of the knowledge transfer. *Collaboration* is the name we give to knowledge transfer between people. Four types of collaboration can take place based on the space and time of the knowledge transfer, illustrated in Table 14.1 (Anumba, Ugwu, Newnham, & Thorpe, 2001): face-to-face collaboration, asynchronous collaboration, distributed synchronous collaboration, and distributed asynchronous collaboration.

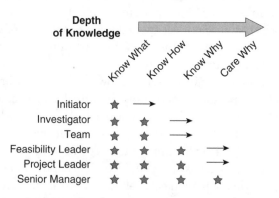

Figure 14.4   Depth of Knowledge for Different Team Roles

Table 14.1   Four Types of Collaboration

	Same Time	Different Time
Same Place	Face-to-Face Collaboration	Asynchronous Collaboration
Different Place	Distributed Synchronous Collaboration	Distributed Asynchronous Collaboration

SOURCE: Adapted from Anumba et al. (2001).

FACE-TO-FACE COLLABORATION

This involves exchanging information at the same time and in the same place. Meetings in a common venue such as a meeting room, office, café, or even corridor are examples of face-to-face collaborations. People are literally looking into each other's faces. An example of this could be a kick-off meeting at the beginning of a project. This type of collaboration is the most common and requires communication skills that improve the sending and receiving of messages.

ASYNCHRONOUS COLLABORATION

This involves exchanging information in the same place but at different times. This can be done using communication media such as physical notice boards. People can view the information displayed in the same place but at different times.

DISTRIBUTED SYNCHRONOUS COLLABORATION

This involves exchanging information in different places but at the same time. This involves real-time exchange of information between people who are located in different geographic areas. Various techniques are available, such as telephones, videoconferencing, and voice over Internet protocol.

DISTRIBUTED ASYNCHRONOUS COLLABORATION

This involves exchanging information at different times and in different places. This can happen when recipients are in different geographic locations but are able to access the same information at different times. This mode of communication involves communication via mail, fax machines, voice mail, pagers, e-mail, and, increasingly, collaborative workspaces and Internet portals.

An organization uses all of these types of collaboration. As teams have become more virtual and people have become more flexible and mobile in task completion, the need for collaboration that is both distributed and asynchronous increases. E-mail, collaborative workspaces, mobile phones, and other electronic tools and techniques are now normal modes of exchanging information in addition to the traditional face-to-face meeting. We call these tools knowledge management tools for convenience.

Knowledge Management Tools

Knowledge management tools support the management of knowledge generation, codification, and sharing in order to enhance decision making. Most tools are computer based, and our discussion will focus on these types of tools. A number of principal technologies are used in creating knowledge management tools and systems and include the following:

- E-mail
- Mobile phones and personal digital assistants
- Web portals
- Content management systems
- Document management systems
- Search engines
- Relational and object databases
- Workflow systems
- Peer-to-peer computer technology
- Intelligent agents
- Customer relationship management
- Data mining and data warehousing

All these technologies are interrelated through a common communication technology and, in most cases, the Internet. It is beyond the scope of this book to review each technology and how it can be used to enhance the innovation process. We focus instead on the use of a simple knowledge management system that can be deployed over the Internet to offer distributed asynchronous collaboration based on innovation-related information. Before we begin looking at the details of such a system, let's review some of the terms and technologies involved.

INTERNET

The Internet is a global network connecting millions of computers, all working together and with the possibility to share information. In the context of knowledge management, the Internet has five main characteristics:

- Cost-effective
- Location and time independent (access anytime, anywhere)
- Distributed connectivity (many computers connected)
- Robust global data path (Internet generally does not fail)
- Global search functionality

By connecting people and organizations all over the world, the Internet accelerates innovation by providing a platform where knowledge and experience can be shared. The Internet distributes information, which can then be shared and transformed into knowledge.

INTRANET

An intranet is a private network of computers that is contained in one organization. It may incorporate many interlinked local area networks, especially in the case of a large extended global organization. Intranets are generally linked to the outside Internet through one or more gateway computers with appropriate firewalls. The main purpose of an intranet is to share company information and explicit knowledge such as documents and policies. Parts of the intranet that are made accessible to customers, partners, suppliers, or other outside organizations become an extranet.

EXTRANETS

An extranet is the part of an intranet that can be viewed by users outside the organization, such as suppliers and customers. An extranet requires a secure transfer of information between the involved partners. The security of the transactions is guaranteed by the firewall server management, the issuance and use of digital certificates or similar means of user authentication, the encryption of messages, and the use of virtual private networks.

INFORMATION PORTALS

These are one or more specific online workspaces such as interactive Web sites or document management systems. Portals can provide services such as online forms, document repositories, document management and version control systems, remote integration, discussion forums, and other tools that support collaborative work. Information portals typically allow a high degree of collaboration between people in a community who share a common purpose. Collaborative portals are an essential component of knowledge management systems that support innovation in organizations.

Collaborative Portals

Distributed asynchronous collaboration systems have become the norm for the sharing of information between those who are in different places and at different times. In many organizations this means everyone. Typical systems include interactive Web sites and internal collaboration portals. Systems and devices such as e-mail and mobile phones can easily be integrated with such systems, allowing a person to be in touch with his or her colleagues anytime and anywhere. Collaborative portals are also known as collaborative workspaces, groupware, and, more generally, knowledge management systems. These terms are essentially interchangeable. *Collaborative portals* is the term we will use in this book.

> *Collaborative portals are computer-based systems used to generate, represent, store, access, and retrieve information in a distributed asynchronous way across an organization.*

Portals typically contain forms, reports, and other codified information to be shared by people in the organization. They can also contain discussion forums, wikis, blogs, bulletin boards, and links to a variety of other resources. They are called portals because they provide organizations with a gateway to large amounts of organizational information. There are two major types: horizontal portals and vertical portals. Horizontal portals serve large communities with a wide variety of information (e.g., Google, AOL, and MSN). Vertical portals serve specific communities and therefore contain domain-specific information. For example, there are vertical specific portals that seek suppliers involved in the car industry and internal enterprise information portals for topics such as innovation management. Our interest in this book is in vertical portals for use in sharing innovation-related information. Some of the specific functions of such portals include the following:

- Simple creation of new forms, fields, and views (i.e., portlets)
- Search and navigation
- Database management
- Document management and version control

- User personalization of interfaces
- Automatic e-mail alerts
- E-mail groups
- Predesigned portlets (e.g., discussion forums)
- Task management and workflow
- Integration with other computers
- User management and authentication control

These functions allow users to share information by uploading and editing information when accessing communal information. Many vendors provide collaborative portals that allow teams to easily configure a portal site for innovation management. We will return to the discussion of collaborative portals in the next chapter when we discuss how to build a simple knowledge management system for managing innovation-related information. For the remainder of this chapter we look at practical issues of the personalization of knowledge: meetings and meeting management.

Example: SharePoint Team Services is a product from Microsoft that is bundled with its main server products. SharePoint provides a portal management environment that allows organizations, teams, and individuals to set up and configure collaborative portals that can then be used to store and share information. This tool allows the user to create a full-featured Web site with built-in functionalities such as announcements, discussion forums, e-mail notification, and address books. The basic components of the system are Web parts (referred to later in this book as portlets). These can be positioned dynamically on the pages that the user has access to. The Web parts allow users to create documents, forms, columns (also known as fields), and views. For example, it is easy to re-create any of the tables presented in the "Activities" section of earlier chapters in SharePoint. A user can create a Web part for "requirements" with a particular form and fields. Later he or she can create various views that display alternative views of the records and their columns. SharePoint can be configured to enable a full-featured information repository for use in any innovation management system. Other tools from different suppliers include Lotus Notes, WelcomeHome, Netscape Collabra, Novell GroupWise, Webflow, and IBM Websphere.

Discussing Results

Face-to-face meetings are one of the most important techniques used in personalization for the purpose of sharing knowledge (e.g., communicating goals and monitoring results). Although meetings undoubtedly form a valuable source of communication, motivation, and understanding, unfortunately they are also a potential waste of time and effort and may also cause individual demotivation. Successful goal definition, project management, and team empowerment depend very much on effective meeting management. Well-managed meetings highlight the inevitable discrepancies and inconsistencies sometimes inherent in innovation plans and also act as a cause for motivation in empowering action. Well-conducted meetings smooth over perceived leadership weaknesses and ill-advised past decisions and help to identify future team leaders. The ultimate goal is to get the most from the meeting in the right time frame while also producing results that are productive, informative, and motivating.

TYPES OF MEETINGS

It typically takes a number of different types of meetings to plan and execute an innovation plan. Some meetings are nonpermanent, occurring only once or twice; others are permanent, recurring repeatedly on an ongoing basis. Permanent meetings are sometimes called workgroups because groups and subgroups of individuals will meet regularly to discuss particular aspects of innovation. Meetings and their behavior will be decided according to a number of factors, including leadership style and the culture of the organization. This section describes typical meetings used in creating and executing an innovation plan. Types of meetings include the innovation workshop, innovation team meeting, steering group meeting, project team meeting, and individual review meeting.

INNOVATION WORKSHOP

This meeting typically takes place once a year and involves all or most members of the innovation team. It also may involve senior managers, consultants, and perhaps an independent facilitator. The purpose of the meeting is to review the previous goals and actions of the organization, discuss the changed organizational environment, and agree on future goals or the suitability of existing goals. The environment review would involve such activities as analyzing stakeholder requirements and reviewing strengths, weaknesses, threats, and opportunities. It would also aid in developing a sense of urgency toward organizational innovation going forward. Setting goals involves agreeing on a vision, setting up strategic objectives and performance indicators for the period ahead, and assigning responsibility for goal management. Initial high-profile actions such as key problems, key ideas, and key projects may also be discussed as possible quick wins to initiate the implementation of the innovation plan. This meeting would usually last for a number of days and would need to be carefully managed, often using a consulting facilitator. It may be necessary to conduct the meeting off-site to avoid the distractions of day-to-day operations. The outcome of the meeting will be a draft innovation plan, later to be fine-tuned and communicated to the organization through more regular innovation team meetings.

INNOVATION TEAM MEETING

This is the most important meeting to be held after innovation goals have been set and takes place periodically, either weekly or monthly. This meeting is usually attended by active members of the organization for the purpose of reviewing goals and agreeing to and reviewing actions. The status of various actions is examined on an ongoing basis, and the various teams are also monitored in order to ascertain their success in attaining objectives. The principal agenda items are a review of goals (i.e., objectives and indicators), discussion of new ideas and projects, a review of existing actions (i.e., problems, ideas, and projects), and a review of the relationships between goals and actions. Not all goals need to be reviewed at each meeting. Short meetings might focus on exceptions such as outliers or urgent goals and actions. Under the discussion of actions, new ideas may be reviewed and decisions made. At this meeting reviews would focus on the portfolios of indicators, ideas, or projects rather than on particular details. Details for each individual goal or action may be communicated verbally by the person responsible.

STEERING GROUP MEETING

At times it may be necessary for a subset of senior staff to meet infrequently to review the overall progress of the strategic aims of an organization. Senior members of the innovation team will arbitrate on strategic issues and monitor key gates in the development of the project portfolio. The group can also be used for resource allocation and budgeting decisions that cannot be resolved in the innovation team meeting. The innovation team may often find that its activities are reviewed by a steering group, which acts as an independent watchdog. In addition, members of the innovation team should understand that their decisions, though not being overruled, may need to be fine-tuned toward the broader goals of the steering group. This is also an opportunity for senior management to subtly impose strategic organizational direction and decision making while preserving the innovation team's autonomy, empowerment, and discretion.

PROJECT TEAM MEETINGS

These take place among the members of individual project teams, focusing exclusively on the goals and actions of one particular project.

Project teams can be a subset of the larger innovation team or a group of suitably skilled people drawn from across the organization. Agenda items usually focus on project tasks, workpackages, deliverables, schedules, resource issues, and any problems that arise. The project leader convenes the meetings and also interprets the decisions of the steering group and innovation team regarding the impact of the project. Individual project team meetings typically contain much more detailed information than that discussed at the innovation team level.

INDIVIDUAL REVIEW MEETING

This meeting takes place between an employee and his or her supervisor regarding a performance review. At the beginning of the year both will have agreed on what specific goals, actions, and teams the employee will engage in. They will also have agreed on what technical, management, personal, and interpersonal skills the employee should be developing. At the end of the year both will review progress toward these objectives and any other developments or contributions the employee has made. The end-of-year meeting is typically preceded by one or two interim meetings during the year to review progress. The outcome of the meeting may be a simple acknowledgment of expectations met or contributions made. The outcome may also lead to some form of reward or recognition of the employee.

EXAMPLE: An innovation team has regular weekly meetings regarding the goals, actions, teamwork, and results of its innovation plan. It has adopted a lightweight approach to recording minutes. The agenda is now fixed in advance of each meeting so each participant knows what items will be covered. The agenda is as follows: (1) previous minutes, (2) actions arising from the minutes, (3) review of goals, (4) review of actions, and (5) any other business. The meetings always begin at 9 A.M. and end at 10 A.M. Actions resulting from previous minutes involve a review of who agreed to do what and whether it was completed successfully. The review of goals concerns exceptional results (i.e., objectives or indicators that are out of control and showing a red light). The person responsible reports on any exceptional objective by reading its highs, lows, and future actions into the record. The purpose of the meeting is to review these highs and lows and consult with this person about all future actions. The same process takes place with the various actions. From time to time nonexceptional goals and actions are reviewed, in particular where some major success or milestone has occurred. The agenda item labeled "any other business" is a catchall for any discussions that arise but do not fit into the regular agenda items. These items are agreed on at the beginning of the meeting and allocated appropriate times by the chairperson.

Summary

Knowledge is information plus judgment that can lead to informed decisions. Most knowledge remains trapped inside the minds of a person or on his or her personal computer. This can be released for use by others through two strategies: codification and personalization. Collaboration is a process that allows people to share knowledge and make decisions that achieve common objectives. There are a number of technologies that support collaboration. In the extended and virtual organization such technologies include e-mail, videoconferencing, and collaborative portals. Collaborative portals are content management systems used for a wide variety of information that can be stored in portlets. There are many potential portlets used for storing and sharing innovation-related information. This chapter has focused mainly on the generic characteristics of knowledge management. In the next chapter we look at designing a simple knowledge management or portal system for an innovation community that uses many of the tables (or portlets) introduced throughout this book.

Activities

This activity requires you to find Web links to at least five software providers of knowledge management systems suitable for replacing the tables used in the various end-of-chapter activities. Search the Internet. Begin by searching for the systems mentioned earlier (Microsoft Share-Point, Lotus Notes, WelcomeHome, Netscape Collabra, Novell Group-Wise, Webflow, and IBM Websphere). Make note of the hyperlinks of the main sites you visit. Copy Table 14.2 into a spreadsheet and complete the fields.

STRETCH: Other elements of this activity may include discovering and testing a free online system that can be used for sharing innovation-related information or project information.

REFLECTIONS

- What is the difference between information and knowledge?
- What are the five stages of the knowledge management process?
- Explain why knowledge sharing is important.
- Explain the difference between tacit and explicit knowledge.

- What does "care why" mean in the context of depths of knowledge?
- Explain "minimum critical codification."
- List a number of technologies that can be used for distributed synchronous collaboration.

Table 14.2   Create Links

Links		
Group	**Title**	**URL**

Group: Label of the group of links (e.g., "Benchmarks" or "Tools")
Title: Title of the link
URL: Link address

Building Communities 15

Previously, we looked at the need to codify and share tacit knowledge about innovation. We also acknowledged the need to create an environment where teams can be empowered to act on the common purpose of innovation. We saw that knowledge management is about supporting the learning process for the individual, the team, and the organization. In this chapter we look at some of the practical or applied aspects of building innovation communities. Two complementary but very different strategies are grouped intentionally in this chapter. The first and most difficult strategy is to build up a learning organization. The ultimate goal of every organization is to become self-sustaining by generating new ideas that eventually become innovations that add value to customers and to create an environment conducive to continuous learning. The second strategy concerns building appropriate information systems for sharing innovation-related information. This is easier than building up a learning organization. Throughout this book we have presented various tables in the end-of-chapter activities. These tables can be converted into simple knowledge management systems. Knowledge management systems assist in creating and establishing routines around the innovation management process and also help to manage the complex amount of innovation-related information. We begin the first half of this chapter by looking at building communities through learning organizations. The second half is dedicated to looking at building a simple information system for supporting innovation data.

LEARNING TARGETS

When you have completed this chapter you will be able to

- Explain the concept of the learning organization
- Understand the individual learning process

- Describe a number of potential sources of knowledge for learning
- Explain the importance of codifying and sharing critical knowledge
- Describe a model for sharing specific innovation knowledge
- Explain the concept of relationships between different sets of data
- Design a simple information system for innovation-related data

Learning Organization

A learning organization is skilled at creating, acquiring, and transferring knowledge. It is also skilled at modifying its behavior so as to reflect new knowledge and insights. This type of organization facilitates the further education of all its members and continually transforms itself through better knowledge and understanding. In the context of the innovation funnel presented earlier, organizational learning is the process by which the organization carries out actions in response to given goals. The results of these actions are then analyzed, and organizational behavior may be changed so as to achieve maximum results by creating new goals and actions if necessary. Organizations may be viewed as engaging in two different learning loops: single and double (Argyris & Schön, 1978). Single-loop learning is that which occurs in the case of an organization examining errors in actions and attempting to correct these errors as they progress. Double-loop learning is the examination of the error occurrence relative to the underlying variables that influenced the action in the first instance. This second learning loop is more reflective and focused on system variables as opposed to individual actions. Considering these two learning loops in the context of the innovation funnel presented earlier, we see that organizations learn from the consequences of particular actions undertaken in response to meeting particular goals and also from the questioning of the appropriateness of the overall framework for achieving the desired future results. Often these two learning loops occur at different levels of the organization and different time intervals.

SINGLE-LOOP LEARNING

In relation to the innovation funnel, single-loop learning occurs in the team implementing a particular action. It occurs on a continuous basis for as long as the action lasts. If the plans or results begin to show a negative result, then the team undertakes analysis to correct the error and begin to incorporate any lessons learned into their future work practices. In situations where a knowledge system exists, the team records and shares this learning experience with other teams across the organization. This continuous learning improves the capability of the organization in its implementation of future innovation actions.

DOUBLE-LOOP LEARNING

In relation to the innovation funnel again, double-loop learning occurs at the senior management or steering group level but on a less frequent basis than the single learning loop. Periodically, the management team will review the operation of the innovation process and its collective output. This analysis is undertaken with a view to enhancing overall efficiency of the process. This review questions the appropriateness of goals and team resources that may constrain the innovation process and the process for the action pathway. By assessing the overall results of the innovation process, managers may decide to change the governing variables for better results. The double-loop learning cycle continues indefinitely as the organization strives to enhance its ability to bring effective innovation actions to fruition.

Developing a Learning Organization

A number of criteria have been identified as beneficial for the development of a learning organization (Garvin, 1993). Learning is recognized as being an important and cyclical process toward more effective innovation. The following criteria have been developed from the perspective of the innovation process:

- An acceptance of the different roles of innovation and operations
- A free flow of information
- The ability to reframe information as far as the innovation perspective is concerned
- The ability to value people as the key element in organizational learning

These conditions reflect many of the discussions in earlier chapters, particularly the discussion of innovation culture and innovation leadership. A learning organization is one that values the act of learning, especially learning from mistakes. Risk taking, experimentation, and benchmarking are all characteristics of such an organization. Ideally it should differentiate between the role of operations and that of innovation. Operations define the process of marketing, manufacturing, and delivering products and services and should continue indefinitely if new and existing customers are satisfied. Innovation is the process of enhancing products, processes, and services to meet the new demands of the market. A learning organization will help its members understand the importance of maintaining this dual focus. The learning organization should give free and open access to information in the areas of learning, analysis, idea generation, problem solving, and goal setting. The final criterion emphasizes that people are at the center of any drive toward creating a learning organization, for no organization can learn without learning by its individual members. Learning is a complex process, and an organization's success is based on the learning potential of its members.

Individual Learning

Individual learning occurs in one of two ways: by individual education and by recruitment of new individuals into the organization (Simon, 1995). Individual learning does not guarantee organizational learning, but without it no organizational learning can occur. Kolb identified the two dimensions of learning style for the individual (Kolb & Fry, 1975): perception and processing.

PERCEPTION

This learning is perceived through concrete experience (feeling) or through abstract conceptualization (thinking).

PROCESSING

This learning is processed through active experimentation (doing) or through reflective observation (watching).

Learning therefore results from individual perception and the appropriate processing of knowledge acquired in this manner. If learning is to be efficient, the learner needs four different types of learning skills: concrete experience, reflective observation, abstract conceptualization, and active experimentation (Kolb & Fry, 1975). Effective learning requires that these learning skills be brought to bear alternately on the learning problem. Therefore, the ideal learning process includes moving continually from concrete experience (feeling), to reflective observation (watching), to abstract conceptualization (thinking), to active experimentation (doing), and then back to concrete experience and repeating the cycle all over again (Figure 15.1).

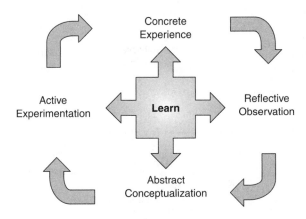

Figure 15.1 Kolb Learning Model

SOURCE: Adapted from Kolb and Fry (1975).

A simpler view of this process is to view every learning activity by a team as consisting of some type of collective reflection; for example, completion of a project should be followed by collective reflection on the highs and lows of implementing it. Without this, the learning process is incomplete, and mistakes on one project may be repeated over and over again by the organization. The resistance to collective reflection can be high in cases where blame or sanction may be imposed for mistakes, yet it is essential if the organization is to develop into a learning organization.

Sources of Learning

An important action in learning is learning by doing. People learn by suggesting ideas, then investigating them. They also learn by implementing projects and reviewing them against stated goals. In this book, readers are encouraged to learn by doing. You read the chapter, and you are then encouraged to complete the end-of-chapter activities so that you can practice some of these lessons. This helps the reflection process and the process of embedding key concepts in memory. Other sources of learning include collecting information. There are many sources of information that provide opportunities for future learning for an organization. Figure 15.2 illustrates some of these. The horizontal axis represents time, and the vertical axis represents the source of the information within the organizational environment.

Sources of learning from the past include experiences of past activities, internal organization reports, and externally sourced books, magazines, and conferences. Sources of learning from the present include internal and external benchmarking, real-time process data, and training programs. Sources of learning from the future include exploration of radical ideas

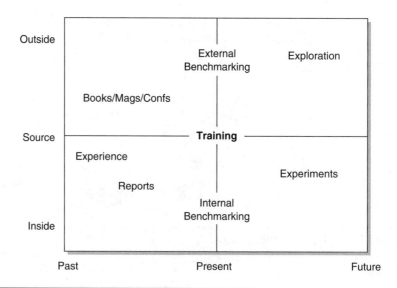

Figure 15.2   Sources of Learning

that may or may not be implemented through experimentation. A learning organization is capable of discerning which learning stimuli it needs to deploy. Systems, policies, and tools then must be put in place to foster the learning process for all those involved. We now turn to the second strategy used for building communities for innovation: creating an information portal for innovation-related information.

Innovation Portal

Creating the learning organization is one way to build communities within the organization. A complementary approach is to build a computer-based collaboration portal. Information systems designed to share innovation-related information must focus on minimum critical codification, that is, only the critical information necessary for implementing goals, actions, teams, and results. It is critical to point out that when we want to share information within a community, we must carefully consider our intended audience. The basic principle is to share information that is of use to the team as a whole and not to clutter up the information system with what is of use to only one or two people. This latter, more detailed information must be treated separately. Focusing on information of use to the team as a whole also avoids wasting valuable time and the creation of communication noise that distracts people from important decisions.

In this part of the book we will refer to the creation of an innovation portal, but as you will see, we will actually be creating something that can just as easily be implemented using a simple spreadsheet. Spreadsheets are extremely useful in this regard. First, they force us to minimize the amount of data we want to codify and later share (i.e., the principle of minimum critical codification), and second, they provide a platform that is widely available and understood. We strongly encourage you to begin the design of your collaboration portal using a simple spreadsheet. Later you will see that these specifications are easily applied to the development of a sophisticated Web-based collaboration portal. We will continue with our use of the term *portal* for the reminder of this chapter.

Throughout the earlier chapters, a number of tables or "information containers" for the innovation process have been presented; these have come under the headings *goals, actions,* and *teams.* Under the heading *goals,* for example, we discussed the need to create statements and indicators. As far as *actions* are concerned, we presented a number of activities that encouraged you to create ideas and projects. All of these and more information containers or portlets are illustrated in Figure 15.3. Portlets are separately designed information containers in a collaborative portal. Many of these stand alone as information subsystems. Others can have relationships with other portlets, and these relationships must be programmed into the portal database.

The reader may notice that all of the portlets in the funnel are expressed as plurals. This is because they generally represent two or more records of information. Think of each portlet as a unique form. Now think of two or more records that use the same form. For example, when we define individuals in an organization, we will be defining two or more individuals with unique names. When we define indicators, we define two or more

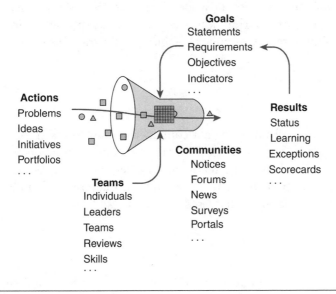

Figure 15.3   Innovation Portal

performance indicators with their own unique titles. The idea that all of these portlets are plurals reinforces the concept that we are essentially managing portfolios of ideas, projects, objectives, and so on. Many other portlets can be added to the model presented in Figure 15.3.

Portal Design

Collaborative portals contain portlets or information containers that provide specific functions. There are many types of portlets; for example, there can be lists of specific records and forms, discussion forums, online surveys, wikis, blogs, and bulletin boards. Teams may configure their own portals to suit the information needs of their own organizations. The technical specification of portals varies widely between vendors and organizations. Before proceeding, it is important to point out that the terms used vary between vendors. One vendor will use the term *portlet,* whereas another will use the term *Web port.* Some vendors will use the term *fields,* whereas other vendors will use the term *columns.* The following features are common across most portals: forms, fields, views, relationships, and hierarchy.

FORMS

Collaborative portals allow users to create forms. Forms are a basic unit of information. For example, they can be created for statements or projects. A typical form contains text, fields, and field labels. When it is filled with unique data, it creates a record. The forms or portlets discussed throughout this book are illustrated in Figure 15.3. Many other portlets can be used for sharing knowledge, such as portlets for sharing presentations, minutes, or policies. Table 15.1 illustrates a simple form and record.

FIELDS

Fields are small units of data contained within forms. Typically they contain alphabetical data, but they can also contain numbers and dates. Unique data can be typed directly into the field or can be selected from a list such as a drop-down menu. Data can also be computed from the values in other fields. Fields are discussed in more detail later.

VIEWS

Views contain a number of records for a particular form or forms. They usually comprise a number of columns that display the labels of the fields and the values of the fields for various records. Many views of the same set

Table 15.1   Sample Forms

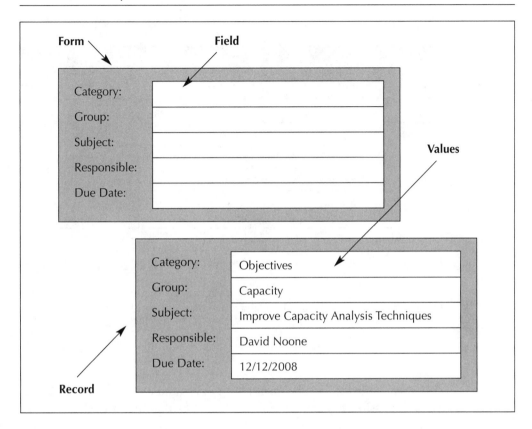

of records can be accessed by simply changing the columns. Views can be configured to filter, sort, and group records according to user-defined criteria (e.g., show only records that have a "red" signal in the status field and sort the records by earliest due date field). Table 15.2 illustrates some of the features of a view.

RELATIONSHIPS

Individual records and lists of records can be configured to have a relationship with other records. For example, we may want to show that a particular project is related to a particular objective. Relationships are discussed in more detail later.

HIERARCHY

Hierarchy is a special type of relationship between records and is usually set up between records in the same list. The parent–child hierarchy is the most common. For example, we may be given a project with a number of

Table 15.2   Sample View

	Field (= Column)			Values	
Objectives					
Group	**Title**			**Responsible**	**Status**
Capacity	Use low-risk strategy for capacity expansion			Mary Roche	☺
Capacity	Improve capacity analysis techniques			David Noone	☹
Capacity	Improve labor flexibility toward capacity changes			Michael Clark	☺
Capacity	Explore make-vs.-buy opportunities			Stewart O'Neill	☺
Responsiveness	Collaborate on development of more accurate forecasts			Danny Mulryan	☺
Responsiveness	Explore manufacture-to-order processes			Michael Clark	☺
Responsiveness	Reduce order delivery times			Stewart O'Neill	☺
Responsiveness	Improve dealer and supplier partnerships			Stewart O'Neill	☹
Organization	Migrate toward flatter and leaner organization			Danny Mulryan	☺
. . .	. . .			. . .	. . .

Category (= Table Title)

Record (= Row)

"children" called tasks. The form for a project and a task may be different, but it is possible to illustrate through hierarchy that they are related.

Field Design

Let us now look at some of the fields that are necessary for sharing innovation-related information. There are a number of aspects to the design of fields; these include the type of field, name of fields, allocation of fields to forms, and additional fields.

TYPES OF FIELDS

There are a number of basic field types, illustrated in Table 15.3. Most fields are simple text fields that contain one string of text (e.g., "Reduce absenteeism within 12 months"). Other common field types include drop-down fields, which give users a choice of key words from which to choose, and number fields, which can contain only number characters (e.g., 234). Another common field is the attachment field. When this field is inserted into the form, it allows users to upload or attach documents to the record.

Table 15.3 Basic Field Types

Field Type	Description
Text	Text field for a single line of text (e.g., 50 characters)
Rich text	Text field for multiple lines of text
Drop-down	Drop-down list of key words
Lookup	Look up key words contained in another list of records
Date	Date (e.g., mm/dd/yyyy, or 10/16/2007)
Number	Numeric characters
Radio	Select one item from a list of possibles (e.g., "yes" or "no")
URL	Special field containing a URL hyperlink
Attachment	Special field to allow attachments to be uploaded to record

NAMES OF FIELDS

Table 15.4 contains a list of all of the fields used in the various activities at the end of each chapter. This table also represents the critical fields necessary for managing information related to the concepts discussed in this book. Many other fields may be added. It is up to each organization to decide which fields to include. Most of these field names are explained in the end-of-chapter activities. One field that does not appear earlier is called "Category." This indicates the name of the form for a particular data set (e.g., indicators or statements).

The status field in Table 15.4 indicates various aspects of the status of an activity. One way to use the status field is to indicate the state of progress of an activity (e.g., "In progress," "Waiting," "Completed"). Another common technique is to add color or graphics to such a field. For example, the "traffic lights" metaphor assigns a meaning to each traffic light color (green, amber, or red) when selected from the drop-down list. Possible meanings for each color are presented in Table 15.5.

Some organizations put a control perspective on the meaning of each color. If the light is red, then the activity is seen as out of control, and the person responsible needs to report on the causes and remedies. This interpretation can sometimes lead the people responsible to hide the true status of the activity because a red signal is equated with personal failure. A softer interpretation is to equate each signal with the need for group discussion. If the light is red, it acts as a signal to discuss progress. In this scenario, the team takes total responsibility for the status of the activity. The person labeled as

Table 15.4 Named Fields

Field	Type	Brief Description
Category	Text	Name of the form (e.g., "Statement," "Indicator")
Group	Text	Group name for the form (e.g., "Thrusts" used on "Objectives" form)
Subject	Text	Name of the record (e.g., "Reduce Absenteeism")
Responsible	Lookup	Select one person from "Individuals" table
Description	Rich text	Description of the record
Status	Drop-down	Select from, e.g., "Red," "Green," "Amber," "Completed"
% Complete	Number	Current percentage completeness of record
Priority	Drop-down	Priority given to the record (e.g., "high," "low," "medium")
Due	Date	Due date for completion of the record
Start	Date	Start date for the start of the record
Private	Radio	Select either "Private" or "Public" for the record
Impact	Drop-down	Impact the record has on innovation (e.g., "high," "low," "medium")
Risk	Drop-down	Risk associated with achieving an impact on innovation (e.g., "high")
Cost	Number	Cost of implementing the record
Payback	Number	Payback associated with achieving goals (e.g., "high," "low")
Link	URL	URL link to more information on the record
Attach File	Special	Link to further electronic attachements associated with the record

responsible takes on the role of watchdog and brings the activity to the attention of the team when necessary. Highlighting the progress of particular actions in this way improves communication within the team.

ALLOCATION OF FIELDS TO FORMS

Not all of the fields presented in Table 15.4 are used in every form or portlet. Table 15.6 illustrates a selection of forms and the fields used in each. Every form includes the "Category" field, which is essentially the name of the portlet or form (e.g., "Statements" or "Indicators"). On the other hand, the field called "Link," which allows users to place a hyperlink in the record, may be present on only one or two forms.

Table 15.5 Traffic Light Metaphor

Color	Interpretation	Alternative Interpretation	Alternative Image
Green	Activity progressing well inside control limits	Activity progressing well	Green smiley face
Amber	Activity entering or leaving "out of control" area	Discussion needed if time is available	Amber neutral face
Red	Activity out of control	Discussion needed	Red unhappy face

ADDITIONAL FIELDS

Many additional fields can be added to the various portlets. Previously we discussed the importance of organizational learning, which allows reflection on the lessons learned during the execution of a particular activity. A number of simple fields can be used for tracking these experiences. During the execution of an activity (goal or action), the person responsible can record a short note on the highs, lows, and future actions. Later, during the innovation team meeting, they can quickly review this information together with the status field for the activity.

▪ Highs are recorded events that went right in the preceding reporting period (e.g., performance gains, increases in participation, good news). Tracking highs gives a historical account of positive things that happened.

▪ Lows are the things that went poorly in the preceding reporting period (e.g., poor performance, disputes, bad outcomes). Tracking and discussing lows is important for knowledge sharing and problem solving within the team and, if possible, avoiding the same lows occurring on other activities in the future.

▪ Future actions are the short-term actions that must be conducted in the next reporting period so that any lows that have occurred in the past may be remedied. They represent attempted solutions to any lows or problems that have occurred. Tracking future actions for a particular activity allows the team to revisit them to assess whether they have been successful.

EXAMPLE: SwitchIt Ltd. is a company that designs and manufactures switch gear. Each form in the innovation team's portal system includes a means to record highs, lows, and future actions to be discussed at the weekly project portfolio meeting. All activities—goals, actions, and teams—have the same fields to allow consistency of communication regarding any lessons to be learned. Figure 15.4 illustrates a truncated project sheet and the fields used by the person responsible for communicating highs, lows, and future actions.

Table 15.6   Fields Used in Various Forms

Fields	Individual	Statement	Requirement	Objective	Indicator	Problem	Idea	Project	Skill	Review
				Selected Forms						
Category	■		■	■	■	■	■	■	■	■
Group		■								
Subject	■	■	■	■	■	■	■	■	■	■
Responsible			■	■	■	■	■	■		■
Description	■	■	■	■	■	■	■	■	■	■
Status			■	■	■	■	■	■		
% Complete			■	■	■	■	■	■		■
Priority						■		■		
Due						■		■		
Start						■		■		■
Private		■		■	■			■		
Impact						■	■	■		
Risk						■		■		
Cost						■		■		
Payback						■		■		
Link									■	
Attach File	■	■	■	■	■	■	■	■	■	■

Relationships

So far we have discussed a variety of information types used in managing innovation in any organization. These have included various types of goals, actions, teams, and results. All of these information sources are important in their own right; however, their importance increases significantly

Project 🔡   ✎ Edit | ✖ Delete

(*)Required Field

Title*	CAD CAM
Responsible	Michael Clark
Team	Andrew Kelly, Danny Mulryan, David Noone, Gary O'Halloran, John Sheehan, Michael Clark, Stewart O'Neill
Description	Find suitable software package to allow transfer of CAD drawings directly onto punching/bending machines.

GOALS

Objectives	Technology
Indicators	Cell manufacturing lead time
Refinements	Reduce lead time
Horizon	Strategic

RESULTS

Status	Green
% Complete	55%
Highs	Sofware and required hardware succesfully installed and set up. Preliminary tests very succesful.
Lows	CNC technicans worried about their jobs.
Actions	Train all manufacturing personnel and the deign team. Send CNC technicans on full course with the software suppliers.

Figure 15.4   Highs, Lows, and Future Actions

when they are related to one another. For example, projects must have goals. Understanding the relationship between projects and goals allows the innovation team to make more informed decisions. In this section, a simple yet powerful technique is presented that will demonstrate the relationships between any two sets of information. Understanding relationships facilitates more informed decisions because it illustrates not only where relationships exist but also where they do not, therefore pointing out a possible gap in the relationship. There are a number of ways to demonstrate the relationship between two or more sets of information, as illustrated in Table 15.7.

The first three types are easy to implement in a knowledge management system; this is done by simply linking one record to others using a drop-down list. The representation of many-to-many relationships is a little more complex but can be achieved using a simple relationship diagram. The diagram illustrates where relationships exist and where they do not. Representation of parent–child relationships can be achieved using a tree diagram. Both of these diagrams are useful for illustrating the many-to-many relationships that exist when comparing two lists. They can also be used to describe the strength of relationships between pairs of data. For the purposes of the activities discussed in this book, a simple relationship diagram offers the functionality necessary for codifying relationships in the innovation funnel.

Table 15.7   Types of Relationships

Relationship	Explanation
One to one	Where one data element is related to another data element (e.g., Initiative 3A is being led by Individual 4B).
One to many	Where one data element is related to a number of other data elements (e.g., Initiative 3A has team members Individual 4B, Individual 5C, and Individual 2E).
Many to one	Where many data elements are related to one data element (e.g., Initiative 3A and Initiative 4D are managed by Individual 4B).
Many to many	Where many data elements are related to many other data elements. This is the same as one to many and many to one but represented together in a matrix.
Parent–child	Where one data element is the parent of another data element. There is a hierarchical relationship between the data elements (e.g., Indicator 2A is a subset or child of Indicator 3S).

The relationship diagram is a decision support tool that facilitates systematic analysis of the relationships between two or more sets of data. Typically, it consists of a table whose first column and top row contain the data lists. The cells that form the matrix between the data lists may contain symbols or numbers that denote the type of relationship that exists between the individual data records. Figure 15.5 illustrates a simple relationship diagram for two data sets, goals and actions; the relationships are denoted by dark dots. It also illustrates an additional row and column for indicating the current results of each record. Relationship diagrams provide a strong visual signal that is simple and easy to interpret. The symbols may be simple dots, or they may be assigned values; these can be summed up to give a numeric indication of the strength of a relationship. The data list in the left-hand column may be interpreted as representing the "what" of a particular analysis; the data list in the horizontal row then represents the "how" in the analysis. For example, "goals" are what need to be achieved and "actions" are how these are going to be achieved.

In general, the data elements belonging to the same row or column should have something in common, so that they make up a set that represents something: a set of objectives or a set of ideas. The strength of the relationship between each pair of data elements is indicated on the cell where they intersect with a symbol (circle, partially shaded circle, or fully shaded circle) or a number (e.g., 1–3, with 3 denoting the strongest relationship). The basic questions that the analyst should ask of the relationships are as follows:

- What is the relationship between any two data elements?
- Why are particular row elements not related to particular column elements?

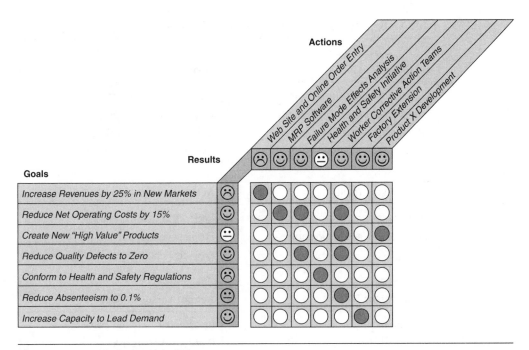

Figure 15.5   Relationship Diagram

- Why are particular column elements not related to particular row elements?

In the case of a diagram of relationships between, for example, objectives and projects, the specific questions would be "Why do some projects have no objectives, and why do some objectives have no projects?" The answers will result in some action being taken: to sanction or discontinue projects that have no alignment with objectives or to create new projects.

TYPES OF RELATIONSHIPS

A number of possible relationship diagrams can be created for all of the activities introduced in earlier chapters and practiced through the various end-of-chapter activities. A number of potential two-dimensional relationships are listed in Table 15.8.

CONSTRUCTING A RELATIONSHIP DIAGRAM

To construct a relationship diagram, the following steps should be taken:

1. Define the purpose of the relationship diagram.
2. Identify the sets of data elements that must be included.

Table 15.8   Types of Two-Dimensional Relationships

Goals Versus Goals	Goals Versus Actions
Objectives versus indicators	Objectives versus projects
Indicators versus requirements	Objectives versus ideas
Objectives versus requirements	Indicators versus projects
. . .	Indicators versus ideas
	. . .
Goals Versus Teams	**Actions Versus Teams**
Objectives versus responsible	Projects versus responsible
Indicators versus responsible	Projects versus team
. . .	Projects versus creator
	. . .
Teams Versus Teams	**Actions Versus Actions**
Individuals versus skills	Ideas versus ranking criteria
Individuals versus courses	Problems versus risk
Leader versus teams	Projects versus schedules
. . .	. . .

3. Assemble a team that can relate all the elements of the relationships.

4. Select the format.

5. Choose and define the relationship symbols.

6. Complete the relationship diagram.

The relationship diagram is a very versatile tool that can be used in many applications. Teams that become "relationship thinkers" gain the ability to conjure up relationship diagrams whenever the need arises, allowing them to explore all available options systematically before making major decisions.

HIERARCHY OF RELATIONSHIPS

Relationships may also have a parent–child relationship. Focusing on a two-dimensional relationship diagram, the row or top list of the diagram can be transposed into the first column of a new diagram. A new list can then be related to these data elements. For example, customer requirements may be related to design features in the first diagram. Design features may then be related to performance indicators in a second diagram, performance indicators related to projects in a third diagram, and so on (Figure 15.6).

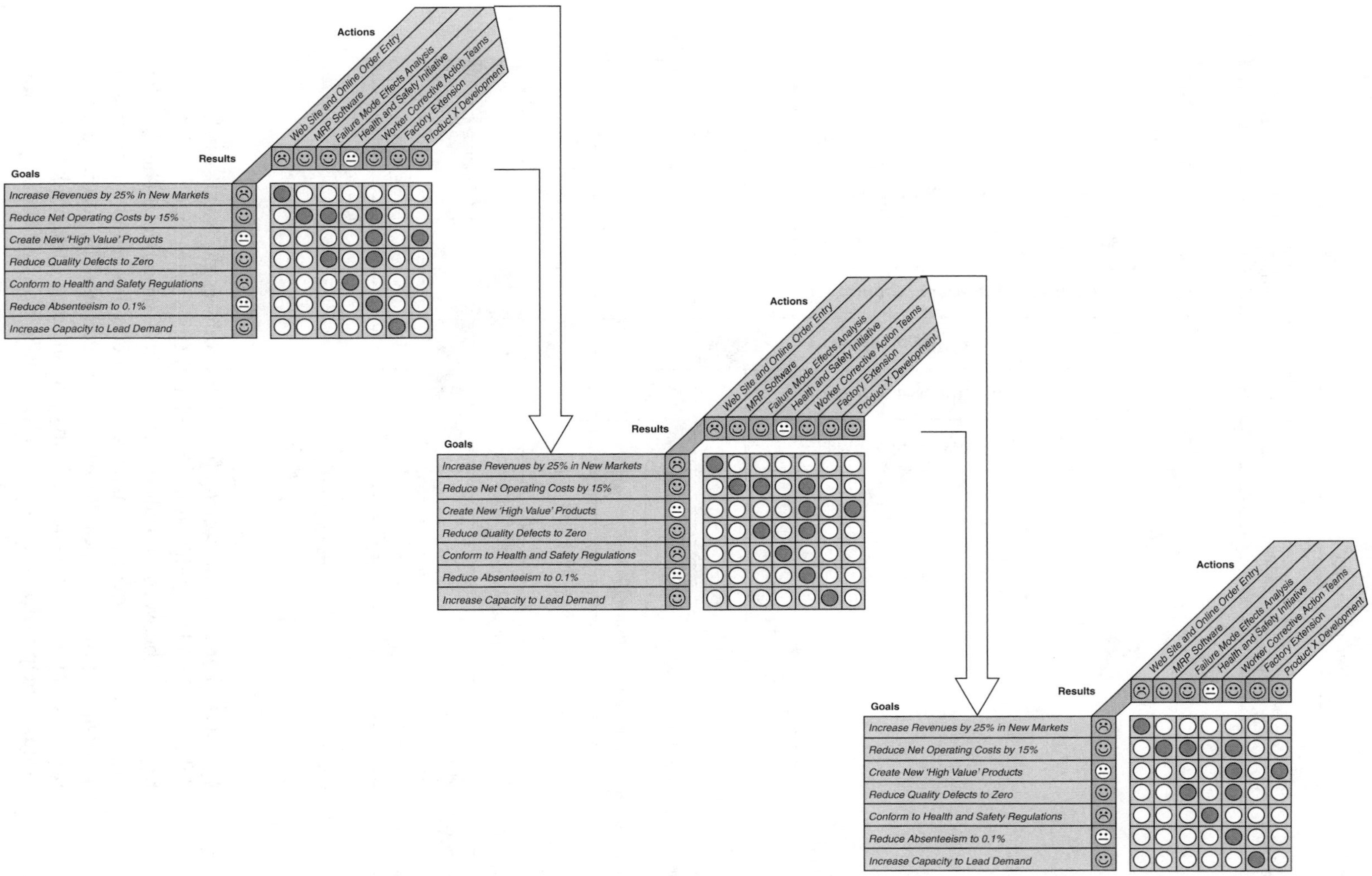
Actions
Web Site and Online Order Entry
MRP Software
Failure Mode Effects Analysis
Health and Safety Initiative
Worker Corrective Action Teams
Factory Extension
Product X Development
Results
Goals
Increase Revenues by 25% in New Markets
Reduce Net Operating Costs by 15%
Create New 'High Value' Products
Reduce Quality Defects to Zero
Conform to Health and Safety Regulations
Reduce Absenteeism to 0.1%
Increase Capacity to Lead Demand
Actions
Web Site and Online Order Entry
MRP Software
Failure Mode Effects Analysis
Health and Safety Initiative
Worker Corrective Action Teams
Factory Extension
Product X Development
Results
Goals
Increase Revenues by 25% in New Markets
Reduce Net Operating Costs by 15%
Create New 'High Value' Products
Reduce Quality Defects to Zero
Conform to Health and Safety Regulations
Reduce Absenteeism to 0.1%
Increase Capacity to Lead Demand
Actions
Web Site and Online Order Entry
MRP Software
Failure Mode Effects Analysis
Health and Safety Initiative
Worker Corrective Action Teams
Factory Extension
Product X Development
Results
Goals
Increase Revenues by 25% in New Markets
Reduce Net Operating Costs by 15%
Create New 'High Value' Products
Reduce Quality Defects to Zero
Conform to Health and Safety Regulations
Reduce Absenteeism to 0.1%
Increase Capacity to Lead Demand

Figure 15.6

Table 15.9   Objectives Versus Projects

Relationships								
	Projects							
Objectives	Install robotic welding	Redesign assembly line	Investigate ERP system	Develop workgroup procedures	Restart sports and social activities	Implement innovation training	Implement e-auctions on selected items	⋮
Use low-risk strategy for capacity expansion		▨	▨	▨				
Improve capacity analysis techniques			▨					
Improve labor flexibility toward capacity changes	▨							
Explore make-vs.-buy opportunities			▨				▨	
Collaborate on development of more accurate forecasts			▨					
Explore manufacture-to-order processes		▨	▨	▨				
Reduce order delivery times	▨	▨					▨	
Improve dealer and supplier partnerships			▨				▨	
. . .								

EXAMPLE: For a manufacturing firm that produces electrical switch gear, Tables 15.9 and 15.10 illustrate the relationships between objectives and projects and between indicators and projects, respectively. The first diagram illustrates where objectives are not being implemented at present and which projects appear to have no objectives. The second diagram does the same for indicators. Management used these and other diagrams to help make decisions about goal definition and action planning.

Table 15.10   Indicators Versus Projects

Relationships								
	Projects							
Indicators	Install robotic welding	Redesign assembly line	Investigate ERP system	Develop workgroup procedures	Restart sports and social activities	Implement innovation training	Implement e-auctions on selected items	⋮
Shipped weight per employee	■		■				■	
Delivery performance	■	■	■				■	
Absenteeism		■		■	■	■		
Defects per unit	■	■		■				
Warranty per 1,000 units per month		■		■				
Manufacturing lead time	■		■					
Cost savings	■	■	■	■	■	■	■	
. . .								

Summary

This chapter has discussed two very different but strongly related strategies to build sustainable innovation communities within organizations: the learning organization and the development of a simple collaboration portal. Learning organizations are adept at learning by their actions, but they also advance through the type of learning that comes from the ambition to do better in the future. The individual is the key to learning in the organization. Collaborative portals facilitate developing and sharing innovation knowledge across the organization. It is possible to specify and build a simple innovation portal that contains a number of forms, fields, and views. A simple system provides more opportunities for communication and personalization of innovation information across the organization. This chapter looked at developing a specification for an easy-to-implement collaboration portal based on tables. The specification contains critical fields, forms, and views that can be used to create routines around the way innovation can be managed. This chapter focused primarily on creating one innovation community with one innovation portal or funnel. In the next chapter we will look at a bigger organization that can contain a large number of communities or innovation funnels. We look at extending innovation beyond a single innovation team and throughout a large extended organization that will contain many departments and innovation teams and may even include stakeholders such as suppliers, distributors, strategic alliances, and user groups.

Activities

This activity requires you to create a number of relationship diagrams for the particular lists in the activities completed in earlier chapters. Begin by creating the following three relationship diagrams: "Objectives Versus Indicators," "Objectives Versus Projects," and "Indicators Versus Projects." Consider other relationship diagrams, such as "Individuals Versus Skills," "Requirements Versus Objectives," and "Activities Versus Projects." Copy Table 15.11 into a spreadsheet and complete the fields.

STRETCH: Other elements of this activity may include creating a number of other relationship diagrams not mentioned here. Consider also finding a suitable online tool for creating and editing simple relationship diagrams.

REFLECTIONS

- Explain organizational learning.
- What is the Kolb model for individual learning?
- Why is reflection important in the learning process?
- Discuss the role of data fields for capturing lessons learned.
- What are the major elements in the design of a simple portal?
- Discuss how relationship diagrams may be used for relationships between different lists of innovation information.

Table 15.11   Create Relationships

Relationships		
	\<List 2\>	
\<List 1\>		

List 1: "Requirements," "Objectives," etc.

List 2: "Individuals," "Indicators," "Objectives," etc.

To paste a list into the \<List 2\> cells, first copy the data you want to paste, then select the first cell in the \<List 2\> cells above, then select Paste Special.

In the dialog box click on the "Transpose" box and then OK.

Now select all cells in \<List 2\> and select Cells from the Format menu.

Select the Format tab and change orientation to 90 degrees. Click OK.

16 Extending Innovation

In previous chapters we explored means for applying innovation within a single community that shares common goals, actions, teams, and results. In most organizations there will be a number of such communities who will deal with various aspects of innovation for products, processes, and services. Within processes alone there may be separate communities for manufacturing, computer services, health and safety, environment, or quality and maintenance. We can visualize each of these communities as a unique funnel with its own goals, actions, teams, and results. Many of these funnels will be directly related to a particular department (e.g., product design, operations, or supply chain). Other funnels will be related to senior management teams that manage a number of departments together. Other funnels may be built around large project teams. Finally, there can be funnels for various strategic partners of the organization such as suppliers, user groups, and distributors. Some funnels will have relationships with other funnels; for example, one funnel could be considered the parent or child of another. Other funnels will deal exclusively with new product ideas, and still more will focus on specific individual projects. The information contained in each of these innovation funnels may be relevant to any person engaged in innovation across the extended organization. This chapter looks at the concept of innovation in an extended organization where there will be many innovation teams, all contributing to growth. We begin by looking at a concept of extended innovation across an organization's network. We then look at a number of different types of innovation processes in the extended organization, from idea repositories up to advanced searching of innovation information. The chapter concludes by presenting a number of simple checklists for assessing innovation in any organization.

When you have completed the chapter you will be able to

- Explain the concept of extended innovation
- Describe a number of different types of innovation processes
- Explain how innovation information can be shared across an extended organization
- Describe a hierarchy of innovation management processes
- Describe future technology for harvesting innovation information
- Describe an approach to presenting and reporting an innovation plan

Extended Innovation

Organizational growth depends on the effective management of innovation. To be effective, innovation must take place in every area of an organization and, by association, in organizational networks that include key suppliers, customers, and other strategic partners. For national and international growth, all organizations—businesses, health services, and even governments—need to engage in innovation. Innovation is needed to improve products and services that add value to customers. It also leads to more efficient processes that help to keep costs competitive, enabling the capacity of more agile organizations to meet unforeseen future challenges. Innovation in an extended enterprise may be visualized through each of the individual funnels of pseudo-independent organizations used by the various innovation teams across the network. Each strategic business unit, department, or even large project of these independent organizations can have its own innovation funnel contributing to some higher-level innovation funnel. If necessary and appropriate, any person in this extended organization may interrogate any of these various funnels across the network in order to gather ideas and information, learn from the experiences of others, and engage in collaborative activities that may lead to even more innovation. Such an extended organization is illustrated schematically in Figure 16.1.

Each of the funnels contains its own goals, actions, teams, and results. Anyone in the extended enterprise may be able to view the innovation activities if security access is granted. For example, a designer in the product design team can visit the manufacturing function's funnel and examine projects that may affect design decisions in the future. A person

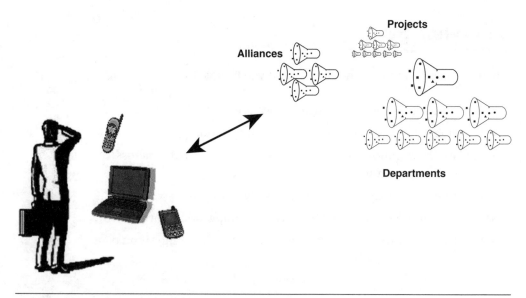

Figure 16.1   Extended Innovation Funnels

in one manufacturing plant may visit the innovation funnel of a sister plant and compare performance indicators. A supply manager in the materials department can visit the innovation funnels of key suppliers and see what actions are being done to improve performance. In this scenario anyone can discover what innovation is taking place anywhere in the extended organization.

The abilities of such an extended organization may be complex, and we will explore clustering later in the chapter. However, the focus for the moment is on the technology for such an innovation management system, which will necessarily combine many different systems and security technologies. One simple system would be one in which all participating organizations simply create their own Web-based portals and agree to place links to their funnels in one master portal for the use of the extended organization. Individuals can then use their Web browsers to click their way around the different funnels. A more advanced approach would be to have all funnels use the same knowledge management software and database, allowing sophisticated searching and sharing of information. A more complex option still is to acknowledge the ubiquitous nature of computing where various approaches and computing technologies are used to deploy an innovation funnel, and use semantic web technologies that allow key information to be harvested from these different systems into one location, where it can then be acted upon by search engines.

Types of Innovation Processes

Throughout this book we have focused on one particular type of innovation management process or funnel that may exist within the boundaries of individual organizations. However, there are other possible types of innovation management processes, and we can divide some of them up according to the number of users they may have. Some key innovation processes and their relationships to each other are illustrated in Figure 16.2. Each has been classified according to the relative number of users and the relative nature of the collaboration. For example, project innovation involves small numbers of people, and the nature of the collaboration is unstructured, so the team can decide precisely how to structure their project in terms of goals, actions, and results. On the other hand, collaborative innovation involves a larger group of people, with more emphasis placed on the use of strict innovation management routines, similar to the one outlined throughout this book. Figure 16.2 illustrates such an innovation hierarchy. Five types of innovation management systems can be identified: individual innovation, project innovation, collaborative innovation, distributed innovation, and clustered innovation.

Individual Innovation

Individual innovation concentrates on the development of effective ways to enhance creative potential from individuals. It has been the focus of

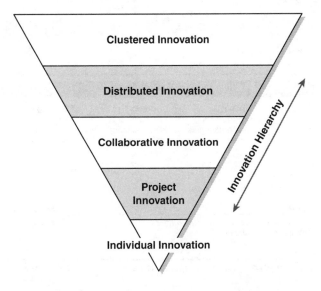

Figure 16.2   Types of Innovation Processes

much research in the past and involves the development of tools that support individuals and small groups engaged in single innovation tasks, such as problem solving and idea generation. Many tools and techniques are available, such as mind mapping, brainstorming, cause–effect diagrams, and Delphi forecasting. These are loosely classified as tools; however, many are more accurately called techniques or methods. These tools may be adapted with varying degrees of success by those involved in the pursuit of innovation. Organizations may find it useful to create a Web-based repository and knowledge source of such tools, which would also act as a repository for ideas and problems. Such a repository portal is illustrated in Figure 16.3. This repository has a number of main functions:

- It contains a comprehensive knowledge base of available tools for users of the system: formalisms, methods, and software tools.

- It contains a repository where ideas and problems can be stored and worked on collaboratively.

- It contains a community area where users of the system can exchange views and opinions.

At this stage of the innovation hierarchy, an individual is encouraged to explore various alternatives for potential innovations using the available tools to kick-start his or her creative capability. This enhances the idea generation and problem-solving capability of the organization.

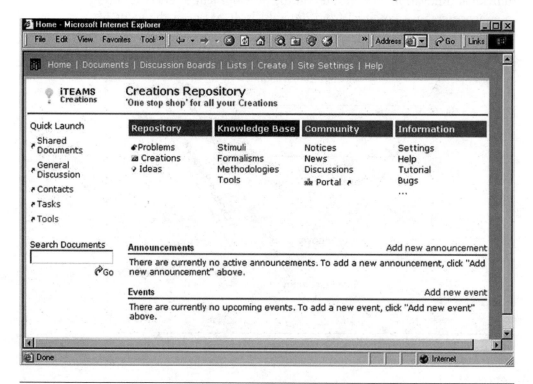

Figure 16.3   Individual Innovation

Project Innovation

Project innovation is concerned with innovation processes that support collaboration across an individual innovation project. The primary community engaged at this innovation level would be the team assigned to the particular project, but in some instances it could encompass participants across an organization. Figure 16.4 illustrates a portal system designed for a particular project. In this tool, all major project activities are codified in a way that follows the method of the innovation funnel discussed earlier. From this figure, it is clear that a number of portlets have been defined and divided into categories of "Goals," "Actions," "Teams," "Results," and "Community." For example, under "Actions" there are a number of portlets for managing project workpackages and tasks, and under "Goals" there are portlets for managing project objectives and project performance indicators. Such project innovation tools are useful in supporting the knowledge requirements of project teams. By using this and other information provided by the project innovation portal, a team can plan and coordinate the development of the projects relative to broader organizational requirements.

Collaborative Innovation

Collaborative innovation focuses on innovation at the organizational level and is typically concerned with portfolios of projects. The organization will also have multiple goals such as strategic objectives and performance indicators. In this configuration, lower-level portals (project portals or funnels) become children of the collaborative innovation portal. The system is designed to be used collaboratively by all system users so that

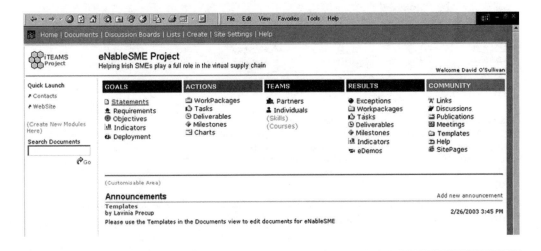

Figure 16.4   Project Innovation

innovation-related information may be shared. It also allows dispersed communities within the organization to collaborate on joint actions, thereby enhancing the sense of common purpose within the community. An example of collaborative innovation is illustrated in Figure 16.5 and reflects the main focus of the activities in this book. By examining the collaborative innovation portal, anyone in the organization will be able to see the portfolio of goals, teams, actions, and ongoing results available.

Distributed Innovation

Distributed innovation is innovation that exists across an extended organization. This level of innovation is defined by all of the collaborative, project, and individual innovation funnels taking place in each of the pseudo-independent organizations within the network. Figure 16.6 illustrates a typical portal for a large corporation, although this could equally represent a network of independent organizations collaborating together. This simple one-page portal contains links to many funnels in the extended organization. Each link takes the user to unique portals or funnels similar to the ones discussed earlier. Users anywhere in the network have the ability to share information, visit partner sites, and benchmark their own performance against similar groups elsewhere in the extended organization. In this way, availability of all innovation portals (individual, project, and collaborative) facilitates network-wide innovation

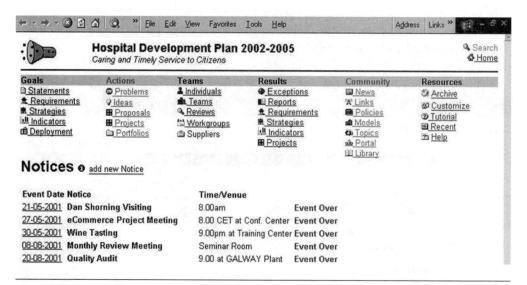

Figure 16.5   Collaborative Innovation

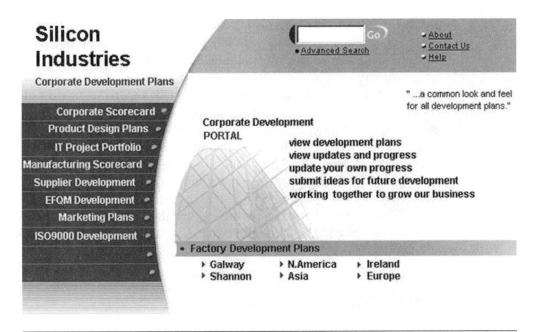

Figure 16.6   Distributed Innovation

by broadening the users' cognitive space and providing links with like-minded users. These kinds of portals are based on the principle of manual browsing, unless all organizations agree to share the same server and database. Although the technical infrastructure addresses certain challenges of distributed innovation, the network must also address more personal issues such as trust, commitment, and skill capacity, all of which are equally fundamental to achieving distributed innovation. We will now deviate slightly from our discussion of types of innovation processes and examine some of the softer and more complex aspects of distributed innovation. We will return to the main thread of the discussion in the section that follows.

Nurturing Distributed Innovation

Organizations often face an innovation paradox: They must innovate in order to compete, but in order to possess the ability to successfully innovate in this area, they may be required to collaborate with other organizations and in some instances competing organizations (Roberts, 2002). Collaborating with pseudo-independent organizations within an extended organization is straightforward because senior managers can dictate the level of collaboration needed to take the organization toward its shared goals. Collaborating with independent organizations and particularly with potentially competing organizations is more complex. The reason for engaging in distributed innovation with independent organizations is primarily necessity. Organizations may undertake distributed innovation with suitable partners to share risk, reduce costs, and access skilled staff or proprietary technology. The benefits of collaboration can lead to knowledge creation, dissemination, exploitation, and learning. Although managing innovation and interaction within the internal boundaries of one organization may be difficult, the challenges become even greater in fostering innovation across a distributed innovation network. These challenges include partnering, collaborative cultures, collaborative trust, physical and cognitive distance, ownership, and partner selection.

PARTNERING

It is very important that suitable partner organizations in the network be organized properly in order to enhance their ability to undertake distributed innovation together. Issues such as the capacity and capability of partners, each individual organization's goals, available innovation budgets, and the ability to collaborate will influence the capability of the network. Any goal or capacity gaps must be confronted at the initial stage of the process to ensure partner compatibility within the distributed network and to ensure effective working relations that will lead to successful innovations.

COLLABORATIVE CULTURES

This is another unique challenge in innovation relates to the knowledge exchange between independent organizations. Traditionally, organizations have been secretive with regard to their innovations in order to protect any emerging intellectual property. Over the years, it has been reinforced in the minds of employees that the business environment is competitive and that others will take competitive advantage if given half a chance. Therefore, the concept of open knowledge exchange with other independent organizations, even within a protected extended innovation network, may be difficult for employees to accept. This may result in apprehension about exchanging knowledge with anyone outside the organization for fear of divulging confidential company information. Organizations collaborating within a distributed innovation network must establish common goals and clear protocols and procedures for knowledge exchange across the network.

COLLABORATIVE TRUST

The presence of trust between collaborating independent organizations is a key determinant in the success of extended innovation. Trust is a psychological state and can be viewed as the undertaking of an uncertain action in the confident expectation that all involved will act honorably and dutifully (Lewis & Weigert, 1985). Trust is a vital component between parties in vulnerable situations because it presumes that one stakeholder will not seek to exploit the other. Trust is not instantaneous within a consortium; it takes time to develop and grow between the parties. Organizations engaging in extended innovation need to ensure that a sufficiently long-term focus exists within the network to allow appropriate trust to develop between the parties. This will increase confidence in the stability of interactions and will reduce the likelihood of opportunistic actions being taken in the short term by any individual partner, to the detriment of long-term benefits for the network. A shared social context should be established, leading to greater levels of trust and understanding between participants as personal relations and mutual respect are built up over time. Issues such as shared social context, various levels of interdependence, and institutional arrangements all contribute to the development of trust within a network and take a significant amount of time to put in place. Well-defined contracts and management structures are often essential at the beginning of the process of collaboration, given the initial absence of prior trust. Parties may resort to control-based reliance (Nooteboom & Six, 2003) until sufficient trust and joint understanding emerge. The provision of suitable management routines, agreed ownership and responsibility, clear project objectives, scope, budgets, and procedures for early intervention when potential difficulties arise can all have a positive impact on the outcome of the collaboration.

PHYSICAL AND COGNITIVE DISTANCE

Another fundamental problem affecting the success of distributed innovation among collaboration organizations is the distance between the stakeholders, particularly geographic and cognitive distance (Balconi, Breschi, & Lissoni, 2002). Physical distance between partners may seriously impede effective knowledge exchange and the success that innovation can bring. Knowledge of a scientific or technical nature may often be tacit, leading to the need for localization and geographic proximity that increases personalization and knowledge exchange. In a distributed innovation scenario, both tacit and explicit knowledge must be exchanged in order for innovation to occur. Recent developments in information and communication technology (including e-mail, intranets, and groupware) mean that geographic distance is becoming less of an impediment for distributed networks. These new developments provide a ready infrastructure for knowledge sharing and the sharing of experience between disparate entities. This increased communication will also enhance social networks between the stakeholders. On the other hand, information and communication technology cannot overcome all the problems of geographic distance between stakeholders because it cannot replace rich communication mechanisms such as face-to-face discussions, practical experiments, and demonstrations, which are often necessary in exchanging tacit knowledge.

A second type of distance that must be reduced between distributed team members is cognitive distance (Balconi et al., 2002). People engaged in distributed innovation can often encounter misunderstandings and disagreements with others in their network through differences in beliefs, behavior, incentives, and motivation across organizations. Organizational culture directly affects the cognitive perspective of organization members. Bringing a distributed community closer together in a cognitive way can often diminish the need for geographic proximity. When managing distributed networks, it is important to reduce the distance between partners by developing appropriate technological and social infrastructure that facilitates knowledge exchange and collaboration.

OWNERSHIP

The ownership of ideas or emerging intellectual property can become a critical issue in distributed innovation. Conflicts of interest over intellectual property rights and opportunism are of no advantage to either party and can damage social capital and trust between them. This could lead to the premature and unsatisfactory termination of the collaboration. In order to facilitate the smooth running of the initiative, the risk exposure of stakeholders and the equitable share of intellectual property rights must be agreed on before collaboration. This would help to clarify the situation as far as intellectual property rights and the revenue streams emerging from them are concerned. This in turn would reduce the likelihood of any conflict of interest or disagreement during the life of the consortium, helping to create a stable environment for collaboration and knowledge exchange. The nature of the collaboration in terms of internal social relationships and the external environment is constantly evolving. Any contractual relationships that have been put in place should be flexible enough to develop with the consortium rather than become a barrier to progress.

PARTNER SELECTION

To deal with the unique challenges posed by distributed innovation, organizations must ensure that they have a number of "soft" factors in place before embarking on a distributed collaboration. The presence of many of these factors will provide the distributed network with a stable foundation on which to undertake distributed innovation. The collaborating organizations can

- Select partner organizations that are compatible with the needs and culture of the network

- Make sure that all partner organizations achieve common understanding of the objectives and implications of the collaboration

- Ensure that clearly defined agreements regarding ownership of intellectual property, division of revenue streams, and roles and responsibilities of partners are in place before collaboration

- Develop suitable technical and social infrastructure to facilitate knowledge exchange by reducing distance between partners

- Ensure long-term focus regarding collaboration in order to allow development of trust and stability between network nodes

Clustered Innovation

Much of this book has focused on the structuring of a knowledge management system for one innovation team that shares common goals. As we saw earlier, there may be many innovation teams within an extended organization. Together they form a single tight-knit community with a common purpose: to stimulate growth for the organization. An innovation cluster, on the other hand, is a loose-knit community of independent organizations. Porter (1998) describes a cluster as an interconnected geographic concentration of companies and institutions in a particular industry sector or field. Such clusters rarely share innovation of a detailed nature across dedicated portals. Rather, they offer information through simple Web sites and a network of relationships, mainly between individuals. The best-known example of an innovation cluster is Silicon Valley, which consists of more than 8,000 independent firms, supported by various other institutions, such as universities, venture capitalists, and development agencies. A number of very large international organizations including Hewlett-Packard, Apple, Intel, Oracle, and eBay are also located there (Lee, Miller, Hancock, & Rowen, 2000).

The key attributes of an innovation cluster include the following (Smith, 2006):

- Geographic concentration
- A high degree of specialization
- Large numbers of small to medium-sized enterprises
- Ease of entry and exit
- High rate of innovation and change

As the cluster develops, its collective innovative capacity increases. This often results from the consequence of people moving between organizations within the cluster, leading to knowledge transfer and the identification of potential synergies that might be exploited. A certain amount of free knowledge exchange within the cluster community is beneficial because it results in greater levels of absorptive capacity and creativity than any one organization could achieve independently. Where synergistic opportunities are identified, these organizations may choose to engage in innovation collaboratively. The cluster innovation relationship is driven by the need for specific innovative actions rather than any long-term holistic strategic alliances between the independent organizations. Therefore, it is possible for one organization to be engaged simultaneously with a number of different organizations in pursuing separate innovative actions. As a result, innovation can occur organically outside the established and semistructured channels of suppliers and strategic partners and wherever the potential for synergistic innovation exists in the cluster community.

Innovation Hierarchy

We now return to the discussion of extended enterprises that contain a large number of pseudo-independent organizations or innovation funnels. It is common for funnels or portals to appear throughout the organization as the building blocks of innovation. Such innovation funnels can be linked hierarchically, but this is often complicated by overlaps (where, for example, two teams share the same results) and by complicated child–parent relationships (where, for example, the output of one project sets the goals of another project). A simplified example of such a hierarchy is illustrated in Figure 16.7. The diagram illustrates a simple hierarchy of innovation funnels, represented as boxes within a large enterprise. They are the building blocks for all the innovation taking place in various departments and strategic business units across an extended organization. Many of these funnels or boxes have hierarchical and peer-to-peer relationships

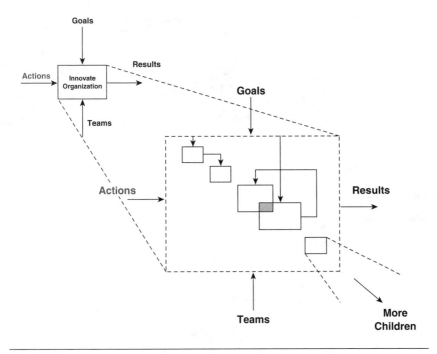

Figure 16.7   Innovation Hierarchy

with each other. Such a hierarchy can be extremely complex where organizations have free rein over how they pursue their goals. On the other hand, they can be simple where there is strong senior management direction and strict routines and protocols for structuring and sharing innovation-related information.

EXAMPLE: Gemini is a group of research institutes that have come together to form an informal alliance. The mission of Gemini is to share information related to common research goals, research projects, research teams, and research results. A number of individual innovation funnels exist within Gemini for each of the participating partners. For example, each research project contains information on project goals, project tasks and deliverables, project teams (individuals and partners), and project results. There are more than 20 research projects shared by each of the partners in the Gemini consortium. By using its own "research institute" funnel, each partner can access its own project innovation funnel and those of its partners. Any person within Gemini, regardless of his or her primary affiliation, can access information throughout the virtual institute. A schematic representation of one branch of funnels within Gemini is illustrated in Figure 16.8.

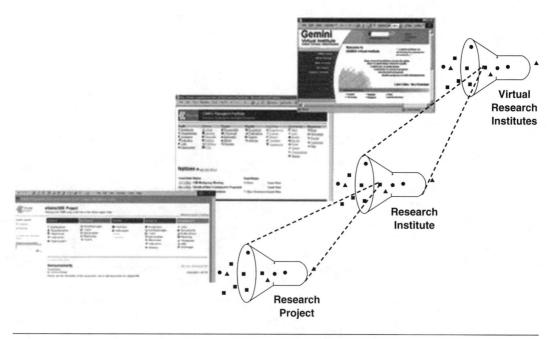

Figure 16.8 Gemini Virtual Institute

Future Portal Technology

Extended enterprises often contain individual organizations that have implemented innovation management systems using a variety of different technologies, software systems, databases, and terminologies. Sharing innovation-related information can be very difficult for people in these extended organizations. One solution is to develop a system that can harvest relevant innovation data from individual systems and map dissimilar terms that have the same meaning. This process is illustrated in Figure 16.9. This system, which is based on semantic web technologies, harvests data from the various innovation funnels over the extended organization. Some of these data are harvested from databases, others from knowledge bases, and still others from Web pages. Information "wrappers" can be programmed to harvest the data regularly. The data are stored in a metadata repository in a form that allows easy search through a "dashboard." A key component of the system is an innovation ontology, or formal description of innovation concepts such as terms and their associations, that allows different formats of innovation data to be mapped and linked together. The innovation "dashboard" allows any user to interrogate the repository for the answer to a large variety of questions about the information stored there (e.g., "What project involving automation has been completed recently?" "Which team or organization is using absenteeism as a

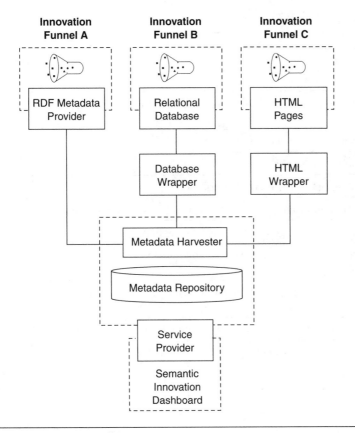

Figure 16.9   Semantic Portal

performance indicator?"). The repository becomes a one-stop shop for all innovation-related information regardless of where it is stored in the extended organization.

Assessing Innovation

We conclude this chapter by looking at some simple checklists for testing our understanding of some of the key issues around managing and applying innovation. The focus is on examining how successful the organization is at practicing some of the concepts introduced throughout this book. Each checklist consists of 10 traits against which the weaknesses or strengths of the organization may be measured (Tables 16.1–16.4). The assessments are divided into the familiar headings of goals, actions, and teams. An organization that scores highly across all checklists may be said to be an innovative organization. This label comes with one critical caveat: that the organization also realizes the potential of all its goals and actions

Table 16.1   Understanding Innovation

	Score (1 to 5)
UNDERSTANDING of the impact of innovation achieved within the organization is clear.	
MANAGEMENT of innovation is effective.	
STRUCTURING of innovation data is effective.	
GOALS of innovation are defined and clearly understood.	
ACTIONS toward creating innovation are managed and clearly linked to goals.	
TEAMS are allocated to achieving goals and performing actions.	
RESULTS of goals, actions, and teams are easily monitored and easily communicated.	
COMMUNITY exists within my organization that is well informed and shares a common purpose.	
RELATIONSHIPS between goals, actions, and teams are effective and transparent.	
COMMUNICATION of ideas and knowledge is fluid and open.	
Total	

Table 16.2   Defining Innovation Goals

	Score (1 to 5)
STATEMENTS for change are concisely defined and documented.	
ENVIRONMENT analysis periodically yields information on strengths and weaknesses.	
BENCHMARKING periodically yields new ideas and comparisons with state of the art.	
STAKEHOLDERS, both internal and external, have been identified.	
REQUIREMENTS of key stakeholders are concisely defined and documented.	
THRUSTS for areas of strategic change and innovation have been identified.	
OBJECTIVES of innovation are defined and foster an improvement culture.	
STANDARDS useful for defining innovation are identified and understood.	
INDICATORS of performance are defined, and targets have been set.	
RELATIONSHIPS between goals and actions are mapped and understood.	
Total	

Table 16.3   Managing Innovation Actions

	Score (1 to 5)
PROBLEMS are effectively identified and solved systematically.	
STIMULI for creativity and idea generation are identified and available for employees.	
IDEAS from employees are welcomed.	
IDEAS are managed effectively.	
PROJECTS are defined concisely.	
PROJECTS are easily ranked according to risk and impact.	
PROJECTS in portfolio are managed using stage gates.	
PROJECTS are classified according to risk and impact.	
RESOURCES are measured and balanced periodically.	
RELATIONSHIPS between goals and actions are mapped and understood.	
Total	

Table 16.4   Empowering Innovation Teams

	Score (1 to 5)
INDIVIDUAL staff members participate in idea generation and problem solving.	
TEAMS, both permanent and nonpermanent, are defined and supported.	
INFORMATION on staff activity is easily accessible.	
REWARD and recognition systems are equitable and transparent.	
REVIEW process links individual performance to company goals.	
REVIEW process allows employees to engage in developing company policy.	
INDIVIDUALS are visibly allocated responsibility for goals and actions.	
INDIVIDUALS have appropriate balance between enablement and empowerment.	
MEETINGS are structured and effective.	
LEARNING culture is provided and promoted within the organization.	
Total	

by adding value to customers and continues to learn from this process so that innovation can be continued into the future. Each list contains 10 traits for an effective innovation management system in any organization. Score yourself on your understanding of each of these traits. Where your understanding is poor, revisit the relevant chapter.

Summary

Many organizations are large enough to have innovation teams that deal with various aspects of product, process, and service innovation. In addition, extended organizations have suppliers, distributors, and other strategic partners that work with the organization in providing products and services to customers. In this chapter, we have seen that many innovation funnels may exist in one extended organization. We have also defined a number of types of innovation systems. We have discovered that it is possible to develop a distributed innovation system that will allow anyone in the system to find innovation-related data. This is used for organizations that have a high degree of control over the technologies and methods being used by the various innovation teams. We have seen how future technologies may integrate different technology platforms to allow advanced search. Technology is perhaps the simplest aspect of building extended innovation. We have also looked at other issues, such as building trust, protection, and shared purpose between different innovation communities that must be addressed to facilitate successful interaction.

Activities

The future plans regarding the innovations that take place in an organization are documented as a matter of course. For example, some of these plans are called development plans or project portfolios. In this book we use the term *innovation plan*. These plans are dynamic. They contain not only the goals set out at the beginning of a period but also the actions (e.g., projects, ideas) that are completed, ongoing, or planned and the teams committed to achieving these actions. Innovation plans are de facto reports on all aspects of the current state of innovation in the organization at a particular time.

Your final activity is to create an innovation plan for your own organization using all of the information gathered in previous activities (see the Appendix for a sample). Review the potential contents of your innovation plan from the samples given. Add text, tables, and figures to your report as you feel necessary.

Undertake a risk assessment of your innovation plan as part of your final report. This assessment can focus on issues such as the current relevance of organizational goals, the mix and status of the action portfolio, the existing skills and competencies available for the process, and the progress of the organization toward its desired future.

STRETCH: Expand your innovation plan beyond the activities presented through this book.

REFLECTIONS

- What is extended innovation in the context of the extended organization?
- What is the difference between collaborative innovation and distributed innovation?
- Explain how hierarchy can be adopted between different innovation funnels in an organization.
- Explain one principal approach that harvests information from distributed information sources.
- What are the main issues to be considered when developing distributed innovation?
- What are innovation clusters?

Appendices

These appendices contain a sample innovation plan for a fictitious organization (Appendix 1) and activities for creating you own innovation plan (Appendix 2). The sample innovation plan, for SwitchIt Manufacturing, also contains a number of tasks that encourage you to expand the plan to incorporate a new design department.

Appendix 2 contains a number of activities that provide a step-by-step guide to creating an innovation plan for your own organization. It gathers together the various activities and tables used throughout the book.

A large number of other innovation plans for various types of organizations are available through the authors' web site (www.owl.ie). Blank and partially completed spreadsheet tables for both appendices are also available from this site.

Appendix 1

Sample Innovation Plan

BACKGROUND

SwitchIt Manufacturing is an Irish manufacturing company and part of the SwitchIt Corporation in the United States. SwitchIt manufactures wall switches and light switches. Marketing is the responsibility of a sister organization based in Brussels, and design is concentrated in the United States. There are 200 employees at the Irish facility. Over the last 15 years the company has built up a mature manufacturing facility for the European and Asian markets. The company is responsible for generating a turnover of €500 million. This year the company is investing €12 million in process innovations, cost improvements, new technology, information system development, and capacity adjustments. A special budget of €1.4 million has also been allocated to establish a new design department in Ireland. The plan in this appendix outlines the status of goals and actions for development of SwitchIt in Ireland over the next 3 years.

TASKS

You have been hired by SwitchIt in Ireland to establish a design department. You have design facilities and have hired one other designer. The managing director has asked that you merge the goals and actions for your new design activity into the plan presented in this appendix. Later, after you have become established, you may create your own separate plan for design activities. SwitchIt manufactures a range of wall switches and light switches. The manufacturing facility has a range of machining and assembly stations that produce the switches in batches according to design specifications. Your role is to extend the product range at SwitchIt by initially using the existing manufacturing processes and skill sets of people involved.

Task: Your first task is to make a portfolio of possible products produced by SwitchIt. Using an Internet search engine, view products from a similar organization. Download pictures and any other information you need. Create a product map showing all the switches in the SwitchIt catalogue.

TEAM

The initial innovation plan illustrated in this appendix was developed by a senior team chaired by the general manager. This team initially met for 1 week off-site to generate the goals of the organization. The team now meets weekly on Fridays for 1 hour to review the status of the company's goals and the status of various actions such as projects and new ideas. This meeting often focuses exclusively on exceptions (i.e., activities that are showing a "red" status signal). In addition to the members shown in Table A1.1, other members of the company are invited to attend as needed.

Task: Add your name and job title and the person you hired to Table A1.1.

Table A1.1   Individuals

Individuals	
Name	**Job Title**
Andrew Kelly	IT Analyst
Brenda Mooney	HR Manager
Danny Mulryan	General Manager
David Noone	Engineering Manager
Gary O'Halloran	Training Manager
James Fogarty	Purchasing Manager
John Sheehan	Quality Coordinator
Mary Roche	Finance Controller
Michael Clark	Manufacturing Supervisor
Stewart O'Neill	Materials Manager
. . .	. . .

MISSION

The mission of SwitchIt is the "Efficient manufacture of innovatively produced switchgear solutions." SwitchIt is focused on manufacturing switch gear at low cost, high productivity, and high quality. It is also focused on continuously improving manufacturing processes. The main contribution to operating revenue and profit is through lowering the total cost of production.

TASK: Propose a change to the mission statement that incorporates the fact that there is a new design department at SwitchIt and a new set of design activities.

ACTIVITIES

A list of manufacturing activities identified in the organization is presented in Table A1.2. The model illustrates that there are three major activities at the top level:

- Manage SwitchIt Ireland
- Plan and Control Manufacturing
- Support Operations

"Manage SwitchIt Ireland" represents the activities of the senior management team in terms of both day-to-day operations and development. "Plan and Control Manufacturing" represents the main activities of producing switch gear in response to customer demand. "Support Operations" includes the activities of finance, engineering, computer services, and human resources.

The "Plan and Control Manufacturing" activity is further divided into three subactivities:

- Plan and Control Materials
- Plan and Control Production
- Ensure and Control Quality

These three primary activities add value to customers and are the focus of much of the development in this current innovation plan. Most goals and actions detailed later focus on these three activities.

TASK: Change the activity model by adding one activity that represents your design department. Remember to use an active verb in the title of the activity. Revisit Chapter 7 to discover how to create a simple activity list.

Table A1.2   Activities

Activities	
Group	**Title**
A0	Operate Switchlt Ireland
A1	Manage Switchlt Ireland
A2	Plan and Control Manufacturing
A21	Plan and Control Materials
A22	Plan and Control Production
A23	Ensure and Control Quality
A3	Support Operations
A31	Provide Personnel Systems
A32	Control Accounting Systems
A33	Provide Engineering Systems
A34	Provide Information Systems

STATEMENTS

As part of the goal generation exercise, a number of statements were initially noted. These include statements of mission and core competencies but also of weaknesses and strengths. Table A1.3 lists a number of these statements, including their status.

The principal weakness is the high rate of product returns through the warranty process due to process-related quality problems. A number of projects are under way to replace some old machines and improve operator training. One of the strengths noted at the beginning of the planning period is the plant's status as a world-class manufacturing facility. Some managers believe that this status is under threat from recent events. Increasing manufacturing costs were initially identified as a potential threat; with rising inflation this appears to be becoming a reality. Finally, one of the opportunities identified at the beginning of the planning period was the availability of university graduates. Because of a number of factors, including the high cost of living, this opportunity may be becoming a threat.

TASK: Add new statements that incorporate the views and analysis of your design department. Visit the Internet and see whether any new technologies are being developed that may offer new opportunities or if

Table A1.3   Statements

Statements		
Group	**Title**	**Status**
Mission	Efficient manufacture of innovatively produced switch gear solutions	. . .
Competencies	Machinists and machining expertise	. . . .
Competencies	Low-tax location and ease of market access	. . .
Strengths	Global organization	☺
Strengths	World-class manufacturing facility	☹
Strengths	Skilled workforce	☺
Strengths	Low employee turnover	😐
Weaknesses	High insurance premium	☺
Weaknesses	Lack of interdepartment communication	😐
Weaknesses	Frequent product returns due to quality problems	☺
Threats	Increasing manufacturing costs	☹
Threats	Competition from new low-cost entrants	☺
Threats	Lack of capital for new projects	😐
Threats	Global downturn continuing	☺
Opportunities	New government design grants	😐
Opportunities	E-commerce opportunities	☺
Opportunities	University graduates	☹

products are being developed by competitors that may constitute a threat. Remember, this is an exercise. If you cannot find potential statements through your research, make them up instead.

BENCHMARKS

Competitors currently lead with market share and are the ones to watch for new product innovations. Two organizations, GE and Philips, are watched closely for signs of new process innovations. The organization also learns much from three associations listed in Table A1.4.

Table A1.4   Benchmarks

Benchmarks	
Group	**URL**
Product	www.lighting-direct.co.uk
Product	www.delixigc.en.china.cn
Product	www.yusinglighting.com
Process	www.ge.com
Process	www.philips.com
Association	www.nema.org
Association	www.sme.org
Association	www.nam.org

TASK: Add other URLs to Table A1.4, particularly for organizations you could watch for new design concepts.

REQUIREMENTS

The company has a number of key stakeholders. Preliminary requirements from these stakeholders are listed in Table A1.5. One of the principal stakeholders is the parent company, which is demanding a €300,000 cost reduction in the current year. Another key group of stakeholders, customers, are requiring shorter lead times, higher quality, and higher reliability. They are also requiring greater flexibility in the event of order changes, with less red tape in changing order quantities and due dates.

Other stakeholders not listed include the design department in the United States and the marketing function in Brussels. The warranty department has also identified a number of requirements, particularly low reliability on some products, that must be incorporated into the plan.

TASK: Add new requirements to this list. What requirements might your parent organization have for your design activity? What new requirements might your customers have for your product portfolio?

Table A1.5   Requirements

Requirements			
Group	**Title**	**Responsible**	**Status**
Parent	Improve cost structure (300k)	Mary Roche	☺
Parent	Greater use of assets	Danny Mulryan	☹
Parent	Pilot corporate ERP systems	Andrew Kelly	☺
Customers	Reduced lead times	Michael Clark	😐
Customers	Increased flexibility	Michael Clark	☺
Customers	Greater quality and reliability	Stewart O'Neill	😐
Employees	Greater discretion and responsibility	Brenda Mooney	☹
Regulations	Health and safety compliance	Luke Davenport	☺
Regulations	Environmental compliance	David Noone	😐
Community	Local sponsorship	Brenda Mooney	☺
Suppliers	Faster payment times	Stewart O'Neill	😐
Suppliers	More accurate forecasting	Stewart O'Neill	😐

OBJECTIVES

The strategic plan adopted by the company at the beginning of the planning period is illustrated in part in Table A1.6. The main decisions for change over the next 3 years have been divided into eight strategic thrusts (or groups): capacity, responsiveness, organization, workforce, supply chain, technology, information, and quality.

Task: Add new objectives to the list in A1.6 that incorporate your design department's objectives. Do you have objectives for new product development? Or perhaps improving reliability of existing products? Or perhaps lowering costs of materials and assembly of existing products?

Table A1.7 illustrates the relationships between objectives and activities (i.e., where changes are needed in the current activity model).

Table A1.6   Objectives

Objectives			
Group	**Title**	**Responsible**	**Status**
Capacity	Use low-risk strategy for capacity expansion	Mary Roche	☺
Capacity	Improve capacity analysis techniques	David Noone	☹
Capacity	Improve labor flexibility for capacity changes	Michael Clark	☺
Capacity	Explore make-vs.-buy opportunities	Stewart O'Neill	😐
Responsiveness	Collaborate on development of more accurate forecasts	Danny Mulryan	☺
Responsiveness	Explore manufacture-to-order processes	Michael Clark	😐
Responsiveness	Reduce order delivery times	Stewart O'Neill	☺
Responsiveness	Improve dealer and supplier partnerships	Stewart O'Neill	☹
Organization	Migrate toward flatter and leaner organization	Danny Mulryan	☺

Task: Add the new objectives and new activities to Table A1.7 and show where relationships may exist.

INDICATORS

The status of key performance indicators is listed in Table A1.8. "Defects per Unit" continues to be a major concern because of a high number of old machines and certain practices among a small number of employees. John Sheehan (quality coordinator) is convinced that machine age and operator training are the main causes of low quality.

Task: Add one or two new indicators to this table that measure the activities of your design department. Remember, this is an exercise.

The key corporate requirement of reducing costs for the Irish operation is a main driver of the innovation plan and relates directly to the indicator of improving cost savings. This indicator is progressing well, as illustrated in the indicator chart, Figure A1.1.

Task: Create a new chart for one of the indicators you defined earlier.

Table A1.7   Relationships (Objectives vs. Activities)

Relationships								
	Activities							
Objectives	Plan & Contol Manufacturing	Plan & Contol Materials	Plan & Contol Production	Ensure & Control Quality	Provide Personnel Systems	Contol Accounting Systems	Provide Engineering Systems	Provide Information Systems
Use low-risk strategy for capacity expansion	■	■						
Improve capacity analysis techniques				■			■	
Improve labor flexibility for capacity changes		■						
Explore make-vs.-buy opportunities	■							■
Collaborate on development of more accurate forecasts		■						
Explore manufacture-to-order processes	■			■				
Reduce order delivery times		■						
Improve dealer and supplier partnerships						■		

Table A1.8   Indicators

Indicators					
Title	**Unit**	**Current**	**Target**	**Responsible**	**Status**
Improve cost savings	€	120k	300k	Mary Roche	☺
Increase delivery performance	%	89%	95%	Michael Clark	😐
Reduce absenteeism	days/month	45	30	Brenda Mooney	☺
Defects per unit	defects/unit	23	10	John Sheehan	☹
Reduce warranty per 1,000 units per month	€	23k	20k	David Noone	☺
Reduce manufacturing lead time	days	5	4.5	Danny Mulryan	😐

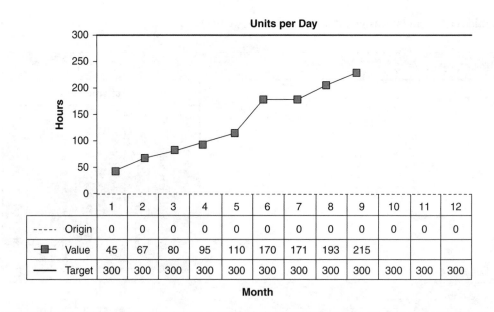

Figure A1.1   Indicator Chart

A strong relationship between objectives and indicators is desirable to ensure that progress can be measured and tracked. As Table A1.9 illustrates, all objectives are linked with the indicators defined.

TASK: Add your new objectives and new indicators to this table and show where relationships may exist.

PROBLEMS

There are more than 230 problems on the reactive problem list, sorted according to impact, risk, and priority. Every machine and assembly station has a proactive problem list with identified tasks for avoiding the problems. Table A1.10 lists a sample of the problems.

TASK: Add some new design-related problems with existing products to this list (i.e., what are the potential problems with switches? Do they overheat? Or does the switch begin to stick after a few months? Or perhaps the cover breaks easily when installed by an electrician.). Remember, this is an exercise; use your imagination to add a few problems to the table.

IDEAS

Every employee is encouraged to generate ideas that can lead to goal attainment. Employees have full access to the objectives and indicators of the organization. Ideas have been grouped by the source of the idea (e.g., goals, problems, new knowledge, benchmarks, employees, customers). Table A1.11 illustrates some sample live ideas.

Table A1.9   Relationships (Objectives vs. Indicators)

Relationships								
				Indicators				
Objectives	Improve Cost Savings	Increase Delivery Performance	Reduce Absenteeism	Reduce Defects per Unit	Reduce Warranty per 1,000 Units per Month	Reduce Manufacturing Lead Time		
Use low-risk strategy for capacity expansion								
Improve capacity analysis techniques								
Improve labor flexibility for capacity changes								
Explore make-vs.-buy opportunities								
Collaborate on development of more accurate forecasts								
Explore manufacture-to-order processes								
Reduce order delivery times								
Improve dealer and supplier partnerships								

TASK: Add some new design-related ideas to this list. Do you want to create a new product? Or perhaps make major adjustments to existing ones? Or perhaps do something radical? Remember, this is an exercise; use your imagination to add a few ideas to the table.

PROJECTS

The top seven approved projects are listed in Tables A1.12 and A1.13. The first table shows the current progress of the projects. The "Investigate ERP System" project is waiting for new information from headquarters. The "Develop Workgroup Procedures" project is also waiting while clarification about participation from key worker representatives is sought.

Table A1.10   Problems

Problems							
Group	**Title**	**Impact**	**Risk**	**Priority**	**Due**	**Responsible**	**Status**
Proactive	Switch housing difficult to assemble	4	3	5	Jun. 07	David Noone	⊡
Proactive	Similar switches interchanged by accident	5	3	5	Jul. 07	David Noone	☺
Proactive	Omission to place decals on subassembly	3	3	5	Jun. 07	John Sheehan	☹
Reactive	Pins shearing during tightening	5	4	5	Apr. 07	Stewart O'Neill	☺

Table A1.11   Ideas

Ideas							
Group	**Title**	**Impact**	**Risk**	**Priority**	**Due**	**Responsible**	**Status**
Goals	Machine replacement program	5	3	4		Stewart O'Neill	⊡
Goals	Staff magazine	2	1	3		Brenda Mooney	☺
Problems	Implement FMEA on machines	4	3	5		David Noone	☹
Knowledge	Lean project management	4	5	4		David Noone	☺

Table A1.13 illustrates the cost–benefit analysis carried out on the current portfolio of projects. There is currently a high priority on the "Install Robotic Welding" and "Redesign Assembly Line" projects.

Task: Add some new design-related projects to this list and try to make one of them radical. Try to make them different from the ideas you created earlier. Do you want to create a new product? Or perhaps make major adjustments to existing ones? Or perhaps do something radical? Remember, this is an exercise; use your imagination to add a few projects to the table.

Table A1.12   Projects

Projects					
Title	**Start**	**Due**	**Responsible**	**% Complete**	**Status**
Install robotic welding	01/03/2006	01/03/2007	David Noone	45%	In progress
Redesign assembly line	01/06/2006	01/12/2006	Michael Clark	35%	In progress
Investigate ERP system	01/03/2006	01/06/2006	Danny Mulryan	90%	Waiting
Develop workgroup procedures	01/01/2006	01/03/2006	Brenda Mooney	30%	Waiting
Restart sports and social activities	01/05/2006	01/09/2006	Brenda Mooney	50%	In progress
Implement innovation training	01/10/2006	01/12/2006	Gary O'Halloran	100%	Complete
Implement e-auctions on selected items	01/04/2006	01/10/2006	Stewart O'Neill	0%	Not started
. . .					

Table A1.13   Project Status

Projects					
Title	**Cost**	**Benefit**	**Impact**	**Risk**	**Priority**
Install robotic welding	120,000	60,000	5	3	5
Redesign assembly line	200,000	10,000	3	1	5
Investigate ERP system	10,000	1,000	4	2	3
Develop workgroup procedures	10,000	2,000	4	5	2
Restart sports and social activities	50,000	5,000	2	2	3
Implement innovation training	25,000	1,200	5	1	4
Implement e-auctions on selected items	120,000	300,000	5	2	4
. . .		. . .			

A bubble diagram illustrating the impact and risk for the current project portfolio is shown in Figure A1.2. The "Restart Sports and Social Activities" project may have a low impact on achieving our overall goals, but the risk is low, and other benefits will accrue in the medium to long term.

Task: Re-create this bubble diagram with your new projects added. Remember! It's an exercise!

The relationship between projects and objectives is illustrated in Table A1.14 in part. It highlights the fact that certain projects have a strong relationship to fulfilling the defined objectives, whereas other projects have no connection.

Task: Add the new objectives and new projects to this table and show where relationships may exist.

The relationship between projects and performance indicators is illustrated in Table A1.15. All projects can be measured through the top seven indicators.

Task: Add the new indicators and new projects to this table and show where relationships may exist.

PROJECT

Details for one of the ongoing projects in the portfolio are shown in Table A1.16. The project has a number of workpackages and associated tasks.

Task: Create a simple list of tasks for one of the radical projects you defined earlier. Try to make the tasks and schedule realistic. Consider market analysis, collaboration with others, patenting, securing venture capital, testing the product in the marketplace, and prototyping.

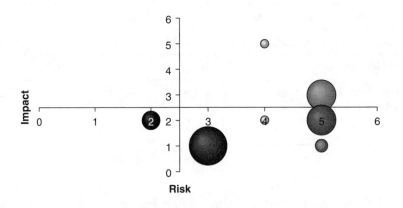

Figure A1.2   Bubble Diagram

Table A1.14   Relationships (Objectives vs. Projects)

Relationships / Objectives	Install Robotic Welding	Redesign Assembly Line	Investigate ERP System	Develop Workgroup Procedures	Restart Sports and Social Activities	Implement Innovation Training	Implement e-Auctions on Selected Items	
Use low-risk strategy for capacity expansion		■	■	■				
Improve capacity analysis techniques	■		■					
Improve labor flexibility for capacity changes			■					
Explore make-vs.-buy opportunities			■				■	
Collaborate on development of more accurate forecasts		■						
Explore manufacture-to-order processes		■		■				
Reduce order delivery times	■						■	
Improve dealer and supplier partnerships			■				■	

SKILLS

The skills or training programs adopted by the team are illustrated in Table A1.17. One new skill has been added this year—"Delegating to Others"—and a customized course for this is being developed by a subcontractor.

TASK: Add a number of skills directly related to the design activity to this list. Visit the Internet and see what courses or learning programs are available to give your team new skills.

Table A1.15   Relationships (Indicators vs. Projects)

Relationships / Indicators	Install Robotic Welding	Redesign Assembly Line	Investigate ERP System	Develop Workgroup Procedures	Restart Sports and Social Activities	Implement Innovation Training	Implement e-Auctions on Selected Items	
Shipped weight per employee								
Delivery performance								
Absenteeism								
Defects per unit								
Warranty per 1,000 units per month								
Manufacturing lead time								
Cost saving								

The relationship between skills and individuals on the team is illustrated in Table A1.18. The dark-shaded cells indicate courses completed and skills acquired by individuals. The light-shaded cells indicate that a course is planned.

Task: Add the new individuals and new skills to this table and show where relationships may exist.

Table A1.16 Project Workpackages

Code	Workpackages and Tasks	Leader	Year 1				Year 2			
			Y1Q1	Y1Q2	Y1Q3	Y1Q4	Y2Q1	Y2Q2	Y2Q3	Y2Q4
WP1	**Project Management**	R1								
T1.1	Establish project, consortium agreement, team portal	R1	■							
T1.2	Manage meetings and goal attainment	R1		■	■	■	■	■		■
T1.3	Undertake progress and cost reporting to the EC	R1		■		■		■		
WP2	**State of the Art and Best Practice**	R2								
T2.1	State of the art in intra/interenterprise constellations	R2	■						■	
T2.2	Best practice in supporting ICTs across industrial sectors	R2	■	■	■					
T2.3	Learning and innovation for virtual teams	R3								
WP3	**Innovation Processes**	R4								
T3.1	Structure and typology in portfolio, program, and project management	R4	■						■	
T3.2	Develop system model for distributed innovation management	R4								■
WP4	**Learning and Innovation**	R3								
T4.1	Model of learning and innovation processes	R3	■							
T4.2	Testing and validation of learning model	R3		■						
T4.3	Development of context-sensitive guidelines	R3			■					

(Continued)

Table A1.16 (Continued)

Code	Workpackages and Tasks	Leader	Year 1				Year 2			
			Y1Q1	Y1Q2	Y1Q3	Y1Q4	Y2Q1	Y2Q2	Y2Q3	Y2Q4
WP5	**Reference Architectures**	R2								
T5.1	Reference architecture requirements definition	R2	■	■						
T5.2	Developing draft reference architecture	R2						■		
T5.3	Validation of reference architecture	R2				■				
WP6	**Ontology and Semantics**	R1								
T6.1	Review of relevant international standards and ontology languages	R1	■	■						
T6.2	Design of an ontology model for innovation	R1				■	■			
WP7	**Prototyping and Design**	R1								
T7.1	Development of tool for program innovation management	R2					■	■	■	
T7.2	Development of innovation learning model	R3					■	■	■	
T7.3	Development of toolset and portals for distributed innovation management	R1					■	■	■	
WP8	**Dissemination and Exploitation**	R1								
T8.1	Organize four regional dissemination workshops across Europe	R2		■				■		
T8.2	Organize a number of focus exploitation workshops	R2			■				■	
T8.3	Develop exploitation plan	R1								

Table A1.17   Training Programs

Skills	
Group	**Title**
Personal	Managing time
Personal	Negotiating skills
Personal	Communication and presentation
Personal	Project management
Interpersonal	Managing conflict
Management	Innovation management
Personal	Leadership
Interpersonal	Teamwork
Interpersonal	Coaching, mentoring, and motivating
Interpersonal	Delegating to others
Interpersonal	Recognizing others and rewarding
Management	Handling pressure and stress management
Management	Planning
Management	Monitoring performance

Table A1.18   Relationships (Individuals vs. Skills)

Relationships							
Individuals	Managing Time	Negotiating Skills	Communication and Presentation	Project Management	Managing Conflict	Innovation Management	Leadership	...
Andrew Kelly	■	■				■		
Brenda Mooney		■		■		■		
Danny Mulryan		■			■		■	
David Noone	■	■						
Gary O'Halloran		■					■	
James Fogarty	■		■		■		■	
John Sheehan		■		■		■		
Mary Roche					■			
. . .								

Appendix 2

Innovation Activities

Applying innovation is a difficult task. The activities suggested here are designed to allow you to develop your own innovation plan for a fictional organization and to struggle with some of the same decisions that an innovation team encounters when managing the innovation process. You can undertake these activities on your own or in a small group. The output of this set of activities is an innovation plan for your own organization—your own case study. For the purpose of these activities, a number of tables are suggested as a means of recording information; you can add fields to these tables or create your own tables as plan requirements demand. All tables are shown at the end of this appendix.

ACTIVITY 1

In this activity you will create some simple pieces of information for your chosen organization. Examples of organizations include hospitals, manufacturing plants, cinemas, software design houses, and even local government agencies. The organization you establish will need to be large. Search online for a real organization that is similar to the organization you are considering developing. Make a note of its homepage address. Research this real organization's innovation effort to discover the goals it pursues, the products and services it has developed for the market, who its competitors are, and what competencies give it competitive advantage. Choose a fictitious name for your organization and decide what products or services you are providing and what markets you are serving. Next, choose a name for your innovation plan that also includes a planning period (typically 1–5 years), such as "Company X Innovation Plan (2007–2010)." Create a simple statement of 5 to 12 words that describes what the current mission of your organization is. Record this information in Table A2.1.

ACTIVITY 2

Record the names and positions in the organizational structure of the key people engaged in innovation in your organization in Table A2.2. These people will be given various responsibilities for the development and implementation of the plan. Choose names and job titles that are credible and realistic.

ACTIVITY 3

Create a list of real competitor organizations that operate in the same sector as your organization. Go online now and identify three to five suitable organizations that you will compete against for market share and survival. Record this information in Table A2.3. As in reality, analysis of these competitor organizations can act as a stimulus for generating ideas for new products, processes, or services or simply provide interesting insights into the way they manage innovation. You may also decide to open a research file on each competitor, to analyze their performance, and track their innovation developments relative to your organization. Now that you have made a start at defining your organization, you need to understand the external macroeconomic and microeconomic pressures acting on the organization.

ACTIVITY 4

Undertake an environmental analysis of the organization using models such as stakeholder analysis, SWOT, PEST, and Porter's 5 Forces. Concisely define your findings in a format similar to Table A2.4.

ACTIVITY 5

Define the various stakeholder requirements being placed on the organization (e.g., customers, corporations, government, and internal success factors). Record these requirements in Table A2.5. In addition, based on your understanding of your organization's situation in its current and future environment and the products or services it is producing, define a concise vision statement (less than 20 words) for the next 3 to 5 years. Record this in Table A2.4. This vision will guide subsequent strategies and performance metrics together with associated innovative actions that the organization will engage in over this period.

Now that your organization has a defined vision statement of where it wants to progress over the defined time period of your plan, you need to define the high-level strategic objectives and performance metrics that will help guide the organization's journey.

ACTIVITY 6

Create a list of twelve or more strategic objectives for your organization, divided into broad strategic thrusts. First, define a minimum of three strategic thrusts, based on the environmental analysis, requirements, and statements you defined in the last several activities. Once you have created appropriate strategic thrusts, define at least two strategic objectives for each. When articulating your objectives, try to keep the number of words to a minimum. Try to define strategic objectives that are general enough to remain relevant for the entire planning period (e.g., 3 years). Avoid strategic objectives that can be achieved in 6 months or less; these may more accurately be defined as projects later. Once you have defined your strategic objectives, assign responsibility for their achievement to one or more members of the organization. Record this information in Table A2.6.

ACTIVITY 7

Create a list of performance indicators for your organization for the calendar year. Select these indicators as measures of the requirements and strategic objectives you have defined for your organization. Specify the origin or start value of the indicator for the start month (e.g., January = 60%) and the target value for the end month (e.g., December = 85%). Each requirement or strategic objective can have one or more indicators, and similarly any specific performance indicator can be linked to the achievement of one or more requirements or objectives. When defining the specific indicator, record the associated origin and target values for that indicator. Try to use an active verb in the title of the indicator (e.g., "Increase," "Decrease," "Maintain"). Also assign responsibility for its achievement to a person in the organization. Record this information in Table A2.7. You should also create one or two sample charts for the indicators you have defined.

By this time, your organization should have an integrated set of goals that will encourage and direct the innovative actions that the organization will engage in over the defined planning period.

ACTIVITY 8

Create a list of about 10 fictitious ideas or solutions to problems for your organization. These creative concepts can come from your environmental analysis, goal definition, benchmarking, or creative capabilities. Some of these ideas may be translated later into projects. Group the creative concepts according to whether they began as a problem or from new knowledge, stakeholder requirements, or other stimuli. Once you have recorded these ideas, assess the suitability of the concepts by scoring their impact and risk on a scale of 1 to 5. Based on this analysis, rank the priority of the ideas using a 1 to 5 scale. Keep the number of words to a minimum. Record this information in Tables A2.8a and A2.8b.

ACTIVITY 9

Over the planning period, some of the problems and ideas investigated and developed will be approved to be implemented as projects. Make a list of ongoing projects to create change in your organization. Record relevant data for each project, such as planned start and end date, project leader, estimated cost, expected return, priority or urgency, and status at that particular time (e.g., "Draft," "Awaiting approval," "On schedule," "Not started yet," "Out of control"). Ensure that there are at least seven projects in your portfolio list. This portfolio can be recorded in Table A2.9. Examine the portfolio spread by creating a bubble diagram.

ACTIVITY 10

Create details for one or two of the projects in your portfolio. Focus on creating a list of workpackages and tasks for each project, together with a set of suitable milestones. In addition to choosing the project leader, define the support team that will facilitate the implementation of the particular project. Create a Gantt chart for the implementation of this project. Record the information in Table A2.10.

ACTIVITY 11

Create a list of skills that the people on your organizational chart possesses that enable them to contribute to the development of innovation actions or that they should possess and record it in Table A2.11.

ACTIVITY 12

Choose one or two individuals from your innovation team and fill out a review form (Table A2.12) for their expected personal goals over the coming year.

ACTIVITY 13

Create a number of relationship diagrams for your innovation plan (e.g., objectives vs. projects, indicators vs. projects, skills vs. individuals). When you have completed these diagrams, revisit each activity and check for accuracy and consistency. Use Table A2.13.

ACTIVITY 14

You now have defined the central elements of your organization's innovation plan. Compile this plan into a document for the senior management team of your organization. Annotate each of the tables you have created with text that describes progress to date, suitability of the portfolio of actions relative to your desired goals, and so on. Discuss the risks in your current plan and make recommendations for any changes that the innovation team will implement. Consider what new tables you want to create. What new fields or columns would you like to create, and what new relationships would be useful to create?

Table A2.1   Create an Organization

Organization	
Name:	
Plan Title:	
Mission:	
Benchmark:	http://
Products and Services:	

Name: Name of your team (e.g., "ABC Corp." or "ABC Corp. Quality Dept.")
Plan Title: Title of the plan (e.g., "Innovation Plan 2007–2010" or "Development Portfolio 2010")
Mission: Mission for your team in about 12 words
Benchmark: Web site of a similar but real organization
Products and Services: List the main products or services offered

Table A2.2   Innovation Team

Individuals	
Name	**Job Title**

Name: Names of the individuals (e.g., John Doe)
Job Title: Job title or skill title (e.g., "Production Supervisor")

Table A2.3   Benchmarks

Benchmarks		
Group	**Title**	**URL**

Group: Label of the group of benchmarks
Title: Title of the link
URL: Link address

Table A2.4   Statements

Statements		
Group	**Title**	**Status**
Mission		
Vision		
Core Value		
Core Value		
Quality		
Safety		
Strengths		
Strengths		
Strengths		
Strengths		
Weaknesses		
Weaknesses		
Weaknesses		
Weaknesses		
Weaknesses		
Opportunities		
Opportunities		
Opportunities		
Threats		
Threats		
Threats		
Threats		
Threats		
Threats		
Threats		

Group: Label of the statement (e.g., "Strengths," "Weaknesses")

Title: The statement in less than 12 words

Status: Status of the requirement (e.g., "Not Started," "In Progress," "Waiting," "Completed")

Table A2.5 Requirements

Requirements			
Group	**Title**	**Responsible**	**Status**

Group: Name of the stakeholder (e.g., "Customer")
Title: Title of the requirement (use the voice of the stakeholder, e.g., "Shorter lead times")
Responsible: Person responsible for reporting the status of the requirement
Status: Status of the requirement (e.g., "Not Started," "In Progress," "Waiting," "Completed")

Table A2.6 Objectives

Objectives			
Group	**Title**	**Responsible**	**Status**

Group: Label of the strategic thrust (e.g., "Technology")
Title: Title of the objective (use an active verb, e.g., "Increase capacity in line with demand")
Responsible: Person responsible for reporting the status of the objective
Status: Status of the objective (e.g., "Not Started," "In Progress," "Waiting," "Completed")

Table A2.7   Indicators

Indicators					
Title	**Unit**	**Current**	**Target**	**Responsible**	**Status**

Title: Title of the indicator (e.g., "Reduce defects per unit")
Unit: Unit of measurement (e.g., hours/day)
Current: Current value of the indicator (e.g., 230)
Target: Target value of the indicator at the end of the planning period (e.g., 260)
Responsible: Person responsible for reporting the status of the indicator
Status: Status of the indicator (e.g., "Not Started," "Red," "Amber," "Green," "Completed")

Table A2.8a   Problems

Problems							
Group	**Title**	**Impact**	**Risk**	**Priority**	**Due**	**Responsible**	**Status**

Group: Title of the problem group (e.g., "Reactive," "Proactive")
Title: Title of the problem
Impact: Impact of the problem on goal attainment from 1 to 5
Risk: Level of risk associated with the problem occurring from 1 to 5
Priority: Priority of the problem from 1 to 5
Due: Due date for completing investigation of the problem
Responsible: Person responsible for investigating the problem
Status: Status of the problem (e.g., "Not Started," "In Progress," "Waiting," "Completed")

Table A2.8b   Ideas

Ideas							
Group	**Title**	**Impact**	**Risk**	**Priority**	**Due**	**Responsible**	**Status**

Group: Title of the idea group (e.g., "Problems," "Suggestions")
Title: Title of the idea
Impact: Impact of the idea on goal attainment from 1 to 5
Risk: Level of risk associated with the idea in achieving its impact from 1 to 5
Priority: Priority of the problem from 1 to 5
Due: Due date for completing investigation of the idea
Responsible: Person responsible for investigating the idea
Status: Status of the idea (e.g., "Not Started," "In Progress," "Waiting," "Completed")

Table A2.9   Projects

Projects										
Title	**Cost**	**Benefit**	**Impact**	**Risk**	**Priority**	**Start**	**Due**	**Responsible**	**% Complete**	**Status**
Project 1										
Project 2										
Project 3										
Project 4										
Project 5										
Project 6										
Project 7										

Title: Title of the project (i.e., replace "Project 1" with your own title, e.g., "Install New Telephone System")
Cost: Cost of the project (e.g., 12K)
Benefit: Annual payback, revenue, or cost avoidance (e.g., 5K or a number from 1 to 5)
Impact: Impact of the project on goal attainment from 1 to 5
Risk: Level of risk associated with the project in achieving its impact from 1 to 5
Priority: Priority of the project from 1 to 5
Start: Start date of the project
Due: Due date of the project
Responsible: Person responsible for leading the project
% Complete: The percentage completeness of the project
Status: Status of the project (e.g., "Not Started," "In Progress," "Waiting," "Completed")

Table A2.10   Project Details

Project:

Code	Title	Resp.	Worker-Months						Year 1				Year 2			
			R1	R2	R3	R4	R5	TOTAL	Y1Q1	Y1Q2	Y1Q3	Y1Q4	Y2Q1	Y2Q2	Y2Q3	Y2Q4
WP1																
T1.1																
T1.2																
T1.3																
WP2																
T2.1																
T2.2																
T2.3																
WP3																
T3.1																
T3.2																
WP4																
T4.1			0	0	0	0	0	0								
T4.2																
T4.3																

Title: Title of the workpackage or task
Resp.: Person responsible for leading the project
Worker-months: Number of worker-months allocated to each workpackage and task
R1: First resource (e.g., person, team, or organization)
Total: Total number of worker-months for each workpackage or task

Table A2.11   Skills

Skills	
Group	**Title**
Technical	
Technical	
Technical	
Technical	
Personal	
Personal	
Personal	
Personal	
Personal	
Interpersonal	
Interpersonal	
Interpersonal	
Interpersonal	
Management	
Management	
Management	

Group: Label of the skill (e.g., "Technical," "Personal," "Interpersonal")
Title: The skill or course in less than 12 words

Table A2.12   Review Form

Individual:		<Individual>	
Supervisor:		<Individual>	
Appraisal Period:		<Month/Year>-<Month/Year>	
Status:			

(during period)

GOALS (start of period)				SCORE (1 to 5) (end of period)
Objectives:	<Objective>			
	<Objective>			
	<Objective>			
Indicators:	<Indicator>			
	<Indicator>			
	<Indicator>			
Skills:	<Skill>			
	<Skill>			
	<Skill>			

Total Score:
(end of period)

Table A2.13   Relationship Matrix

Relationships		
	<List 2>	
<List 1>		

List 1: Requirements, Objectives, etc.
List 2: Individuals, Indicators, Objectives, etc.
To paste a list into the <List 2> cells, first copy the data you want to paste, then select the first cell in the <List 2> cells above, then select Paste Special.
In the dialog box click on the Transpose box and then OK.
Now select all of the cells in <List 2> and select Cells from the Format menu.
Select the Format tab and change orientation to 90 degrees. Click OK.

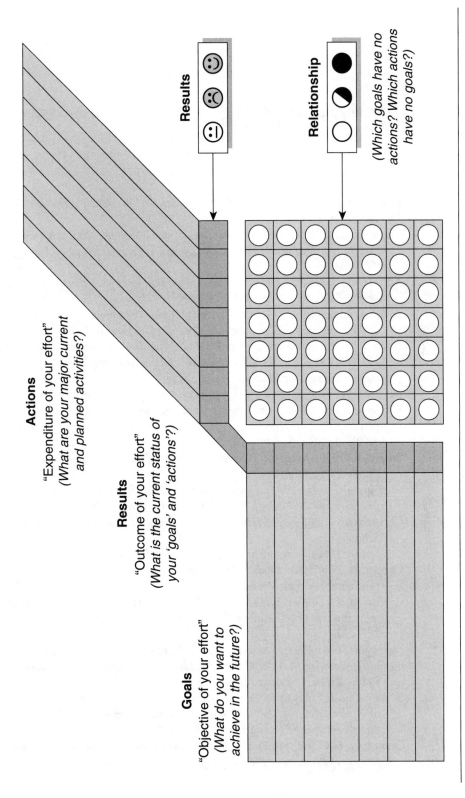

Figure A2.1 Simplified Planning Tool

References

Amabile, T. M. (1998, September/October). How to kill creativity. *Harvard Business Review,* pp. 77–87.

Anumba, C. J., Ugwu, O. O., Newnham, L., & Thorpe, A. (2001). A multi-agent system for distributed collaborative design. *Logistics Information Management, 14*(5/6), 355–367.

Argyris, C., & Schön, D. (1978). *Organizational learning: A theory of action perspective.* Reading, MA: Addison Wesley.

Bacon, F. R., & Butler, T. W. (1998). *Achieving planned innovation: A proven system for creating successful new products and services.* New York: Free Press.

Balconi, M., Breschi, S., & Lissoni, F. (2002). *Networks of inventors and the role of university research: An exploration of Italian data.* http://econpapers.hhs.se/paper/cricespri/wp127.htm

Bessant, J., & Tidd, J. (2007). *Innovation and entrepreneurship.* New York: Wiley.

Burnes, B. (1991). Managerial competence and new technology: Don't shoot the piano player, he's doing his best. *Behaviour and Information Technology, 10*(2), 91–109.

Burnes, B. (1996). *Managing change: A strategic approach to organisational dynamics* (2nd ed.). London: Pitman.

Chesbrough, W. H. (2003). *Open innovation: The new imperative for creating and profiting from technology.* Boston: Harvard Business School Press.

Christensen, C. M. (1997). *The innovator's dilemma.* Boston: Harvard Business School Press.

Cooper, B. (1986). *Winning at new products.* London: Kogan Page.

Cooper, R. G. (2000). *Product leadership: Creating and launching superior new products.* New York: Perseus.

Cormican, K., Dooley, L., O'Sullivan, D., & Wreath, S. (2000). Supporting systems innovation. *The International Journal of Innovation Management, 4*(3), 277–299.

Davenport, T. H. (1992). *Process innovation.* Boston: Harvard Business School Press.

Davila, T., Epstein, M. J., & Shelton, R. (2006). *Making innovation work.* Upper Saddle River, NJ: Wharton School Press.

Dooley, L., & O'Sullivan, D. (1999). Decision support system for the management of systems change. *The International Journal of Technological Innovation and Entrepreneurship (Technovation), 19,* 483–493.

Dooley, L., & O'Sullivan, D. (2000). Systems innovation manager. *International Journal of Production, Planning and Control, 11*(2), 369–379.

Dooley, L., & O'Sullivan, D. (2001). Structuring systems innovation: A conceptual model and implementation methodology. *Journal of Enterprise and Innovation Management Studies, 2*(3), 177–194.

Dooley, L., & O'Sullivan, D. (2003). Developing a software infrastructure to support systemic innovation through effective management. *The International Journal of Technological Innovation and Entrepreneurship (Technovation), 23*(8), 689–704.

Doz, Y., & Hamel, G. (1998). *Alliance advantage: The art of creating value through partnering.* Boston: Harvard Business School Press.

Drucker, P. (1988, November/December). The discipline of innovation. *Harvard Business Review,* pp. 149–157.

European Commission. (1996). *Mapping innovation in Europe.* Brussels: Author.

Flynn, M., O'Sullivan, D., & Dooley, L. (2003). Idea generation within the context of organisational development. *The International Journal of Innovation Management, 7*(4), 417–442.

Franklin, C. (2003). *Why innovations fail: Hard-won lessons for business.* London: Spiro.

Garvin, D. (1993, July/August). Building a learning organization. *Harvard Business Review,* pp. 78–91.

Goffin, K., & Mitchell, R. (2005). *Innovation management: Strategy and implementation using the pentathlon framework.* London: Palgrave Macmillan.

Hackman, J. R. (2002). *Leading teams: Setting the stage for great performances.* Boston: Harvard Business School Press.

Hamel, G., & Prahalad, C. K. (1990, May/June). The core competence of the corporation. *Harvard Business Review, 68*(3), 79–93.

Handy, C. (1985). *Understanding organizations.* Harmondsworth, UK: Penguin.

Hansen, M. T., Nohria, N., & Tierney, T. (1999, March/April). What's your strategy for managing knowledge? *Harvard Business Review,* pp. 106–116.

Hauser, J. R., & Clausing, D. (1988, May/June). The House of Quality. *Harvard Business Review, 3,* 63–73.

Hayes, R. H., & Wheelwright, S. C. (1984). *Restoring our competitive edge: Competing through manufacturing.* New York: Wiley.

Hayes, R. H., Wheelwright, S. C., & Clark, K. (1988). *Dynamic manufacturing: Creating the learning organization.* New York: Free Press.

Henderson, R. (2007). *Developing and managing a successful technology and product strategy.* MIT Lecture Notes.

Johnson, G., & Scholes, K. (1997). *Exploring corporate strategy: Text and cases* (4th ed.). London: Prentice Hall.

Johnson, G., & Scholes, K. (2002). *Exploring corporate strategy: Text and cases* (6th ed.). London: Prentice Hall.

Johnson, P., Heimann, V., & O'Neill, K. (2001). The "wonderland" of virtual teams. *Journal of Workspace Learning: Employee Counseling Today, 13*(1), 24–30.

Jones, P. (1996). *Delivering exceptional performance: How to align the potential of organizations, teams, and individuals.* London: Financial Times Publishing.

Kao, J. J. (1989). *Entrepreneurship, creativity and organization: Text, cases, and readings.* Upper Saddle River, NJ: Prentice Hall.

Kaplan, R. S., & Norton, D. P. (1996, January/February). Using the balanced scorecard as a strategic management system. *Harvard Business Review,* pp. 75–85.

Katzenback, J. R., & Smith, D. K. (1993). *The wisdom of teams: Creating the high performance organization.* Boston: Harvard Business School Press.

Kelley, J., & Littman, T. (2001). *The art of innovation.* Grand Haven, MI: Brilliance Audio.

Kim, W. C., & Mauborgne, R. (2000, September/October). Knowing a winning business idea when you see one. *Harvard Business Review,* pp. 129–136.

King, N., & Anderson, N. (1995). *Innovation and change in organizations.* London: Routledge.

Kolb, D., & Fry, R. (1975). *Towards an applied theory of experimental learning.* New York: Wiley.

Kotter, J. P. (1990). *A force for change: How leadership differs from management.* New York: Free Press.

Kotter, J. P. (1996, March–April). Leading change: Why transformation efforts fail. *Harvard Business Review,* pp. 59–66.

Lampikoski, K., & Emden, J. B. (1996). *Igniting innovation: Inspiring organizations by managing creativity.* Chichester, UK: Wiley.

Lawler, E. (1992). *The ultimate advantage.* San Francisco: Jossey-Bass.

Lee, C. M., Miller, W. F., Hancock, M. G., & Rowen, H. S. (2000). *The Silicon Valley edge.* Stanford, CA: Stanford University Press.

Lewis, J. D., & Weigert, B. B. (1985). Trust as a social reality. *Social Forces, 46,* 967–985.

Light, D. W., & Lexchin, J. (2003). Will lower drug prices jeopardise drug research: A policy fact sheet. *The American Journal of Bioethics, 4*(1), W1–W4.

Luecke, R. (2003). *Managing creativity and innovation (Harvard Business Essentials).* Boston: Harvard Business School Press.

Luecke, R. (2004). *Creating teams with an edge: A complete skill set to build powerful and influential teams (Harvard Business Essentials).* Boston: Harvard Business School Press.

Luecke, R. (2006). *Performance management: Measure and improve the effectiveness of your employees (Harvard Business Essentials).* Boston: Harvard Business School Press.

MacLaughlin, I. (1999). *Creative technological change: The shaping of technology and organizations.* London: Routledge.

Martin, M. J. C. (1994). *Managing innovation and entrepreneurship in technology based firms.* New York: Wiley.

Maslow, A. (1954). *Motivation and personality.* New York: McGraw-Hill.

Maslow, A. (1962). *Towards the psychology of being.* Princeton, NJ: Van Nostrand.

McGrath, M. E. (1996). *Setting the PACE in product development: A guide to product and cycle-time excellence.* Newton, MA: Butterworth-Heinemann.

McGregor, D. (1960). *The human side of enterprise.* New York: McGraw-Hill.

Meyer, J. P., & Allen, N. J. (1990). The measurement and antecedents of affective, continuance and normative commitment to the organization. *Journal of Occupational Psychology, 63*(1), 1–18.

Mintzberg, H., Quinn, J. B., & James, R. M. (1988). *The strategy process: Concepts, contexts and case studies.* London: Prentice Hall.

Moore, J. A. (1999). *Crossing the chasm: Marketing and selling technology products to mainstream customers* (2nd ed.). New York: HarperCollins.

Nadler, D. A., & Tushman, M. L. (2004). Beyond the charismatic leader: Leadership and organisational change. In R. Katz (Ed.), *The human side of managing technological innovation* (pp. 258–272). Oxford, UK: Oxford University Press.

Neely, A., Adams, C., & Kennerley, M. (2002). *The performance prism*. London: Prentice Hall.

The New Oxford Dictionary of English. (1998). New York: Oxford University Press.

Nonaka, I., Toyama, R., & Byosiere, P. (2001). A theory of organizational knowledge creation: Understanding the dynamic process of creating knowledge. In M. Dierkes, A. B. Antal, J. Child, & I. Nonaka (Eds.), *Handbook of organizational learning and knowledge* (pp. 491–517). Oxford, UK: Oxford University Press.

Nooteboom, B., & Six, F. (2003). *The trust process in organisations*. Cheltenham, UK: Edward Elgar.

Olson, E. M., Walker, O. C., Jr., & Ruekert, R. W. (1995). Organizing for effective new product development: The moderating role of product innovativeness. *Journal of Marketing, 59*, 48–62.

O'Reilly, C. A., III, & Tushman, M. L. (2004). The ambidextrous organization. *Harvard Business Review, 82*(4), 74–81.

O'Sullivan, D. (2002). Framework for managing business development in the networked organisation. *Computers in Industry, 47*, 77–88.

Pava, C. (1983). *Managing new office technology: An organizational strategy*. New York: Free Press.

Peters, T., & Waterman, H. (1988). *In search of excellence: Lessons from America's best run companies*. New York: Grand Central.

Porter, M. (1980). *Competitive strategy: Techniques for analyzing industries and competitors*. New York: Free Press.

Porter, M. E. (1998, November/December). Clusters and the new economics of competition. *Harvard Business Review*, pp. 77–90.

Rafii, F., & Kampas, P. J. (2002). How to identify your enemies before they destroy you. *Harvard Business Review, 80*(11), 115–123.

Robbins, S. (1998). *Organisational behaviour: Concepts, controversies, applications* (8th ed.). Upper Saddle River, NJ: Prentice Hall.

Roberts, E. M. (1988). Managing invention and innovation. *Research-Technology Management, 31*(1), 11–29.

Roberts, E. (2002). *Innovation: Driving product, process and market change*. San Francisco: Jossey-Bass.

Rogers, E. M. (1983). *Diffusion of innovation* (3rd ed.). New York: Free Press.

Rosenfeld, R., & Servo, J. C. (1991). Facilitating change in large organisations. In J. Henry & D. Walker (Eds.), *Managing innovation* (pp. 28–38). London: Sage.

Rothwell, R. (1992). Successful industrial innovation: Critical success factors for the 1990s. *R&D Management, 22*(3), 221–239.

Rothwell, R. (1994). Towards the fifth-generation innovation process. *International Marketing Review, 11*(1), 7–32.

Schein, E. (2004). *Organizational culture and leadership* (3rd ed.). San Francisco: Pfeiffer.

Schilling, M. (2006). *Strategic management of technological innovation*. New York: McGraw-Hill.

Scholtes, P., Joiner, B., & Streibel, B. (1996). *The team handbook*. Madison, WI: Oriel Incorporated.

Schumpeter, J. A. (1942). *Capitalism, socialism and democracy*. New York: Harper.

Shapiro, S. M. (2001). *24/7 innovation: A blueprint to surviving and thriving in an age of change*. New York: McGraw-Hill.

Sheth, J. N., & Ram, S. (1987). *Bringing innovation to market: How to break corporate and customer barriers.* New York: Wiley.

Simon, H. A. (1995). Rounded rationality and organizational learning. In M. D. Cohen & L. S. Sproull (Eds.), *Organisational learning* (pp. 175–187). London: Sage.

Smith, D. (2006). *Exploring innovation.* London: McGraw-Hill Education.

Smith, P. G., & Reinertsen, G. (1995). *Developing products in half the time.* New York: Van Nostrand Reinhold.

Stevens, G. A., & Burley, J. (2003, March/April). Piloting the rocket of radical innovation. *Research-Technology Management,* pp. 16–25.

Stewart, T. A. (1997). *Intellectual capital: The new wealth of organizations.* London: Doubleday.

Strebel, P. (1999). Why innovation fails. In *Harvard Business Review on change* (pp. 139–157). Boston: Harvard Business Press.

Takeuchi, H., & Nonaka, I. (1986, January/February). The new new product development game. *Harvard Business Review, 64*(1), 137–146.

Tidd, J., Bessant, J., & Pavitt, K. (1997). *Managing innovation: Integrating technological, market and organizational change.* New York: Wiley.

Tidd, J., Bessant, J., & Pavitt, K. (2005). *Managing innovation: Integrating technological, market and organizational change* (3rd ed.). New York: Wiley.

Trott, P. (2005). *Innovation management and new product development* (3rd ed.). Upper Saddle River, NJ: Prentice Hall.

Tuckman, B. (1965). Developmental sequence in small groups. *Psychological Bulletin, 63,* 384–399.

Utterback, J. M. (1996). *Mastering the dynamics of innovation.* Boston: Harvard Business School Press.

von Hippel, E. (1994). *The sources of innovation.* New York: Oxford University Press.

von Stamm, B. (2003). *Managing innovation, design and creativity.* New York: Wiley.

Wallas, G. (1926). *The art of thought.* New York: Harcourt Brace.

Wenger, E., McDermott, R., & Snyder, M. (2002). *A guide to managing knowledge: Cultivating communities of practice.* Boston: Harvard Business School Press.

West, M. A., & Farr, J. (1990). *Innovation and creativity at work.* Chichester, UK: Wiley.

Wheelwright, S. C., & Clark, K. B. (1992). *Revolutionizing product development: Quantum leaps in speed, efficiency and quality.* New York: Free Press.

Wield, D. (1986). Organisational strategies and practices for innovation. In R. Roy & D. Wield (Eds.), *Product design and technological innovation* (pp. 147–165). Milton Keynes, UK: Open University Press.

Index

About the Authors

David O'Sullivan, PhD, is a senior academic and researcher at the National University of Ireland, Galway. He is also director of research; he and his team study the application of computer technology to the management of innovation in organizations. David also works with multinational and small to medium-sized enterprises on various aspects of learning and research into innovation management. Before joining NUI Galway he was a product design engineer and later a system design engineer with a number of multinational corporations. Details of his work can be found on his Web site, www.owl.ie.

Lawrence Dooley, PhD, is a college lecturer in enterprise and innovation in the Department of Management and Marketing, University College Cork, Ireland. His research interests focus on the management of organizational innovation and related issues of interenterprise collaboration and university–industry knowledge exchange.